SpringerBriefs in Molecular Science

History of Chemistry

Series Editor

Seth C. Rasmussen, Department of Chemistry and Biochemistry, North Dakota State University, Fargo, North Dakota, USA

Springer Briefs in Molecular Science: History of Chemistry presents concise summaries of historical topics covering all aspects of chemistry, alchemy, and chemical technology. The aim of the series is to provide volumes that are of broad interest to the chemical community, while still retaining a high level of historical scholarship such that they are of interest to both chemists and science historians.

Featuring compact volumes of 50 to 125 pages, the series acts as a venue between articles published in the historical journals and full historical monographs or books.

Typical topics might include:

- An overview or review of an important historical topic of broad interest
- Biographies of prominent scientists, alchemists, or chemical practitioners
- New historical research of interest to the chemical community

Briefs allow authors to present their ideas and readers to absorb them with minimal time investment. Briefs are published as part of Springer's eBook collection, with millions of users worldwide. In addition, Briefs are available for individual print and electronic purchase. Briefs are characterized by fast, global electronic dissemination, standard publishing contracts, easy-to-use manuscript preparation and formatting guidelines, and expedited production schedules. Both solicited and unsolicited manuscripts are considered for publication in this series.

Peter Spellane

Chemical and Petroleum Industries at Newtown Creek

History and Technology

 Springer

Peter Spellane
Department of Chemistry
New York City College of Technology,
The City University of New York
Brooklyn, NY, USA

ISSN 2191-5407 ISSN 2191-5415 (electronic)
SpringerBriefs in Molecular Science
ISSN 2212-991X
History of Chemistry
ISBN 978-3-031-09628-0 ISBN 978-3-031-09629-7 (eBook)
https://doi.org/10.1007/978-3-031-09629-7

This Springer imprint is published by the registered company Springer Nature Switzerland AG
The registered company address is: Gewerbestrasse 11, 6330 Cham, Switzerland

This book is for Joanne and for Catherine.

*The universe is made of stories,
not of atoms.*

*Muriel Rukeyser, from "The Speed of
Darkness"*

Acknowledgments

I became aware of the early chemical industries in and near New York harbor when I joined a group of City Tech colleagues on a project supported by the National Endowment for the Humanities. "Water and Work" inspired us to see and learn from our city's waterfront. But it was the period maps at the New York Public Library, the Lionel Pincus and Princess Firyal Map Division, and the Map Division's digital versions of early maps, that fascinated me. The maps revealed the proximities of production sites to one another and to railways, canals, and harbors; maps from different periods revealed the evolution and advances in the several industries at Newtown Creek. The maps suggested an economic geography that made me want to understand New York City's history of production of chemicals. As I began to understand New York's history in chemistry, I joined meetings of the History of Chemistry division of the American Chemical Society. I gathered stories and details from collections at libraries and archives. One story led to another; one library led to another. I wish to acknowledge the libraries and librarians and archives and archivists at which and with whom I worked as this project got under way. These include the Briscoe Center at the University of Texas at Austin, the Brooklyn Historical Society, the Brooklyn Public Library, the Brown University Library, the Cooper Union Library and Archive, the Division of Old Records at the Office of the New York County Clerk, the Geography and Map Division of the Library of Congress, the Library at the Graduate Center of the City University of New York, the Library at the Herreshoff Museum in Bristol, Rhode Island, the Library at New York City College of Technology CUNY, the Mystic and Noank Library, the Mystic Seaport Museum Library, the New-York Historical Society Library, the New York Public Library, the NYPL's Map Division and the Science, Industry, and Business Library, the New York University Library, the Queens Public Library, the Rockefeller Archive at Pocantico, New York, and the Yale University Library and Archive.

Contents

About the Author

Peter Spellane teaches organic chemistry at New York City College of Technology, a campus of the City University of New York (CUNY). He had studied at Hamilton College, the University of Washington, and the University of California at Santa Barbara. Before joining CUNY, he had worked as a research chemist at Akzo Nobel Chemicals in Dobbs Ferry, New York and Arnhem, the Netherlands.

Abstract

Advances in methods of production and growth in volumes of materials produced along Newtown Creek in the second half of the nineteenth century had profound consequence for the practice of industrial chemistry in the United States, for the economic vitality of the City of New York, and for the site's ecology. This book reconstructs Newtown Creek's industrial expansion during the period that began in the 1840s and continued through the early decades of the twentieth century: in that period, production of reagent chemicals and refined materials grew as practitioners, alert to new advances in chemical science, developed and applied increasingly sophisticated production technologies. Industrial practice progressed from the recovery of animal tissues to the refining of petroleum and the production of high-purity metals from mineral ores. With attention to each company's technical expertise and principal products, this book examines the interdependence of the chemical- and material-producing industries that took root and thrived along Newtown Creek's industrial shores.

Newtown Creek's history includes stories of well-known New Yorkers—Peter Cooper, Charles Pratt, John D. and William Rockefeller—and other less celebrated or less notorious characters. The city itself plays a role in the Newtown Creek story. New companies identified and registered themselves at the office of the New York County Clerk, secured capital in the city's financial markets, and constructed innovative production facilities along the shores of New York's inner harbor. Along those waterfronts, barge-loads of raw materials were transformed by chemistry into fine fuels and metals, often sealed in proprietary containers, and packed on railcars or ships for markets in all parts of the country and corners of the world. New chemistry-based ventures thrived on New York City's workforce, its ambition, its capital, and its inclination for expansion.

Chapter 1
Newtown Creek and New York City

This book examines the sequence of manufacturing businesses that began operations along Newtown Creek in the nineteenth and early twentieth centuries. All produced materials of one kind or another; nearly all made use of original and often proprietary chemical technologies.

Newtown Creek is a narrow waterway at the eastern edge of New York harbor. In the second half of the nineteenth century, the unpretentious creek was a most successful incubator space and a bustling center of industrial production. Several of the chemicals-manufacturing and petroleum-refining businesses that began their practice at Newtown Creek became industrial archetypes. Nichols Chemical Company[1] was one such venture. Other businesses, including Standard Oil, that had begun operations at various inland sites, relocated some or all their refining or manufacturing operations to Newtown Creek. Standard Oil secured land near and along Newtown Creek just as the company was chartered in Cleveland. Several of the Newtown Creek start-ups grew into early versions of modern Fortune 100 companies. The success of the Newtown Creek companies, the breadth, originality, and quality of their product lines, provides a fundamental model of the vigor and originality of modern industrial production in the United States. The site's earliest manufacturing industries were glue-making and fertilizer-production. As industrial production matured at Newtown Creek, the site became home to companies producing concentrated sulfuric acid, refined petroleum products, and electrolytically-refined copper. Indeed, refined and valuable material recovered from earth-abundant (although increasingly difficult to extract) raw matter became Newtown Creek's characteristic product-descriptor.

After the discovery of subterranean pools of crude oil in western Pennsylvania in 1859, petroleum refining became the site's major industry. Refining requires reagent chemicals ("reagents"), chemical compositions that effect a change in a substrate

---

[1] Nichols Chemical evolved through a merger of several companies to become the General Chemical Company and, in a later merger of five companies, Allied Chemical & Dye Corporation, which together produced acids, alkalis, coal tar, and aniline dyes, the fundamental product line of chemicals manufacturers in the early twentieth century.

P. Spellane, *Chemical and Petroleum Industries at Newtown Creek*,
History of Chemistry, https://doi.org/10.1007/978-3-031-09629-7_1

material. Crude petroleum is refined in two processes; the first is fractionation, the process of isolating the various components (fractions) of crude petroleum. Fractionation is achieved by distillation: smaller (lower molecular mass) molecules form gases at lower temperature than do larger (higher molecular mass) compounds. The second step in petroleum refining involves treating the separate fractions with reagents that enable the removal of contaminants from each valuable fraction. The ties between petroleum refining and chemical manufacturing are strong; indeed, the one-hand-washes-the-other relationship of reagent chemicals and refined petroleum was, to some degree, forged at Newtown Creek. The impressive growth in scale and sophistication of the two industries reflects both the practitioners' creativity and the bare-knuckle competition that characterized the country's early industrialization and enabled both remarkable technological advances and staggering accumulations of wealth. This was the age of the industrial robber barons.

Questions of how and why these industries thrived in that challengingly navigable corner of New York harbor, an actual "backwater," and the larger question of how industrial zones assemble themselves and why they succeed, led me to write this book. My first question was, "Why chemicals and petroleum?" Neither seemed in any sense "native" to the cosmopolitan waterfront city, a place far from oilfields and mining operations and, I naively thought, from markets for concentrated sulfuric acid or high-purity copper. I wondered about the relationships of the producers among themselves and the succession of the different companies. A second question about Newtown Creek's industrial history concerned location: why did this vigorous industrialization take place in New York, and why along Newtown Creek? One imagines New Yorkers, even of that period, to prefer trading real property and financial obligations to the skinning and tanning of horsehides. How is it that businesses that require dissolving the bones of horses in sulfuric acid or roasting minerals mined in Arizona prospered so well in the center of New York harbor? Why did these rough and polluting industries thrive in New York and remain there until their success demanded more acreage and waterfront than Newtown Creek could provide?

The density of population in and around New York and the vigor and efficiency of the harbor provide answers to questions of "why chemicals?" and "why New York?" Before chemicals-manufacturing and petroleum-refining began at Newtown Creek, the site was home to industries that had more to do with farming than with fuel. Before industries related to agriculture established themselves at Newtown Creek, the site supported fishing and whaling.[2] The area's earliest manufacturing industries recovered value from the carcasses of horses: fertilizer and glue, refined from animal tissues, were Newtown Creek's first large-scale industrial products.

New York has never been a typical American city. New York is a city of islands; only one the city's five boroughs, the Bronx, is part of the North American mainland. Real property in New York and particularly in Manhattan Island has always been precious; land value alone would make the city an unlikely venue for production of

---

[2] The New York-based whaleboat owners A. C. and D. C. Kingsland ran whaling ships from Newtown Creek. Early maps of the area indicate the "Kingsland Docks" and "Whale Creek" on the southern shore of Newtown Creek.

materials, fuels, and metals. But the city's commercial assets are indisputable: New York City and its several islands lie at the eastern edge of a bountiful continent. New York's harbor is ideal: deep water surrounding a cluster of islands, small, medium, and large, all protected from the Atlantic Ocean by two bays that form a miles-long, deep-water channel. European explorers, sailing into the uncharted harbor in the seventeenth century, must have been astounded to discover a long, well-forested island in that harbor and astounded again on realizing that the river that had carried them from the Atlantic Ocean to Manhattan Island continued to the north, beyond the northwestern edge of the island for more than 300 miles. Indeed, the North River would provide passage to an even greater wilderness, a place the Europeans would have to acknowledge to be a new world.

The city dates from about 1624 when Peter Minuit (c. 1580–1638) established an outpost for the Dutch West India Company. Legend holds that the Europeans acquired the island by barter with the people native to the area, glass beads for real property, and images of the West India Company's settlement at the island's southern tip suggest a quaint Dutch village, a community protected from the wilderness to the North by a wall at Wall Street. The outpost was named, with irony or ambition, New Amsterdam. Had the name of the Dutch village survived the subsequent imperial conquest, it would have been apt. Like Amsterdam, New York has the cosmopolitan character of a center of trade and commerce and the feel of a city at the water's edge. The Dutch West India Company's control of the village lasted until in 1664, when Governor Peter Stuyvesant (c. 1592–1672) ceded the island to an English naval squadron. The English renamed the city to honor the Duke of York.

As the English colonists in New York grew weary of their King's financing his European ventures with taxes collected from his American subjects, those subjects nurtured an identity as persons endowed with rights and articulated that identity in letters of protest written in New York. Before the colonies were "states," and well before the young nation's Constitution was signed, the Stamp Act Congress, the assembly that would decry taxation without representation, met at New York's City Hall. So too, a few years later, would the Congress of the Confederation, the nation's first central government under the Articles of Confederation, and, later still, the Continental Congress met there. After the American revolutionary war and establishment of the Constitution of the United States, the first U.S. Congress met at the same location; the building was rechristened "Federal Hall." When the young nation elected its first president, President George Washington swore his oath of office on the balcony of the same Federal Hall. Indeed, in the early years of the country's Federal Period, despite both Boston's and Philadelphia's leading roles in securing the country's independence, New York was venue for country's governance. As the United States matured, its government moved to a new location, a city constructed specifically for the new nation's governance, at a riverside site nearer to the geographic center of the young nation.

Having executed its role in launching the new nation, New York seized upon the entrepreneurial opportunities enabled by the natural assets of the land and the nation's independence to pursue new ventures in manufacturing and trade. With governance of the nation removed from Wall Street, New York directed its energy to the work that became its hallmark: making, selling, and trading. Indeed, in that period, the City of New York, at the southern end of Manhattan Island, evolved: having been a place of political debate, New York grew to become the center of trade that the modern world associates with the name "Wall Street."

At the beginning of the nineteenth century, New York had several fundamental assets of a center of manufacturing and trade: New York, the point of entry for thousands of immigrants, had the country's largest urban population. Many, mostly European, immigrants settled in New York; many arrived with training and competence acquired in their native countries. Many, one assumes, would maintain ties to their old-world contacts.

Manhattan and the harbor that surrounds it seem designed for commerce and sea-borne trade: Manhattan is a long and narrow island that rests upon three layers of bedrock, Manhattan Schist, Inwood Marble, and Fordham Gneiss [1], a 22-mile-long pile of rock, two miles wide at its widest. The North (Hudson) River flows along the length of Manhattan's western side; the East River and the Harlem River flow along the island's eastern and northeastern shores. At Manhattan's northeastern corner, the East and the Harlem Rivers meet at an acute angle. The tidal straits flow together at a watery site called "Hell Gate." The confluence flows to the east past islands named Randall's and Riker's and into the Long Island Sound, a broad flow of the Atlantic Ocean that washes the southern shore of Connecticut and the northern shore of Long Island. At its eastern end, the Long Island Sound opens into the Atlantic Ocean.

The southern tip of Manhattan Island, the site of the original Dutch settlement, is separated from the Atlantic Ocean by the Upper and Lower Bays of New York's outer harbor. To reach the Atlantic Ocean from New York, ships sail to the south and east, through the Narrows between Brooklyn and Staten Island and pass Sandy Hook roughly four miles south of Manhattan. The broad North River that flows along the western shore of Manhattan provides navigable passage 200 miles to the North and, after completion of the Erie Canal in the early nineteenth century, navigable connection to the Great Lakes.[3] The Erie Canal connected New York City to what would then be thought the western United States.

As can be seen in the 1818 map of the United States given in Fig. 1.1, the area west of the Great Lakes, which appear near the top center of the United States and form part of the border with Canada, is labeled North West Territory. In contrast to the detail indicated within New York State, the map, which predates the Erie Canal by a few years, shows no evidence of development to the west of the Great Lakes. On its completion, a few years after the map in Fig. 1.1 was made, the Erie Canal provided a water route for transport of persons and goods between New York and the North West Territory.

Figure 1.2 reveals the fundamental features of the City's harbor: its protection from the open Ocean by the miles-long New York Bay and its river access on both

---

[3] The Great Lakes of North America comprise a set of five interconnected freshwater lakes in the upper Northeast of the United States. The Great Lakes connect to the Atlantic Ocean by the St. Lawrence River, which flows between northern New York State and Ontario, Canada, and connects, by various canals, to the Hudson River and New York City.

Fig. 1.1 Map of the United States of America, 1818 (Courtesy of The New York Public Library [2])

East and West sides of the island. Note that Newtown Creek is indicated in this early map: across the East River from the center part of Manhattan island.

New York secured its standing as the leading port in North America in the decades that followed the wars that secured the independence of the United States, the Revolutionary War and the War of 1812. As the new nation achieved greater stability, its productivity grew, and foreign demand for American products increased. Trade through New York flourished. At the same time, the City's population soared. Ships that had sailed to England with American exports returned with European immigrants. Most arrived in New York harbor, and many went no further than their port of entry.

The trade advantages that nature provided New York, its edge-of-the-continent location and its ample harbor, were leveraged by two inventions that transformed American practice in international trade: the Erie Canal and the packet ships.

Construction of the Erie Canal would enable water-borne commerce between ports on the Great Lakes and along the Canal route with New York City. The institution of "packet lines," cargo-bearing ships that received all manner of items for transport to distant markets, sailing at scheduled departure dates and times, would facilitate commerce between New York and European ports. In time packet lines would link

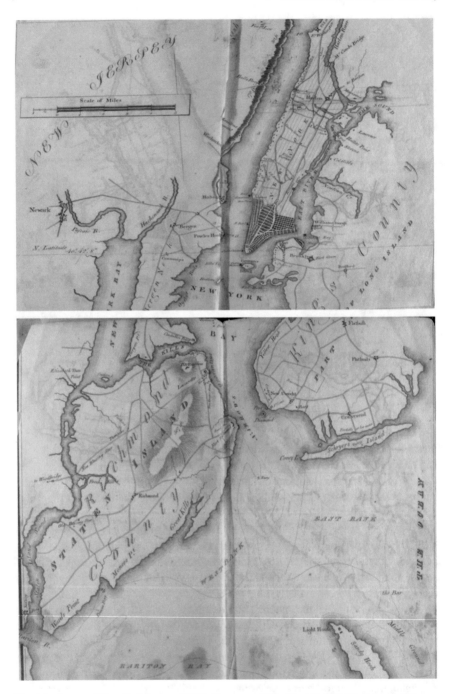

Fig. 1.2 A map of New York harbor (1820) (Courtesy of The New York Public Library [3])

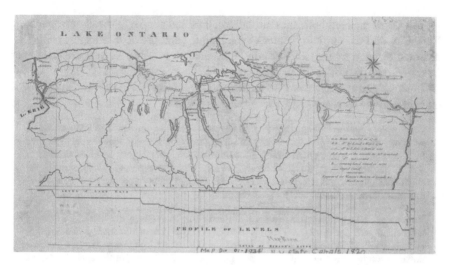

Fig. 1.3 A map indicating the route and water elevations of the Erie Canal (1820) (Courtesy of The New York Public Library [4])

New York to English and continental ports, to the "cotton ports" in the southern United States (Savannah, Charleston, Mobile, and New Orleans), and to harbors serving the New England states. The Erie Canal and the packet lines were ideas conceived in New York, both in or about 1817, with New York harbor in mind. The two concepts, the planned construction of inland waterways and the scheduled departures of packet ships, quickly found application in places and ports far from New York.

The Erie Canal, the water route and series of locks that would provide a navigable connection between ports on the Great Lakes and cities along the route of the canal (Rochester, Toronto, Buffalo, Cleveland, Detroit, Chicago, Milwaukee, and Duluth) and New York City, was envisioned by DeWitt Clinton (1769–1828). The route of the Erie Canal, locations of locks, and water levels along the route are indicated in a period map, Fig. 1.3 [4]. In studying the history of the Erie Canal, a researcher could be forgiven for thinking that the Erie Canal was not only envisioned by DeWitt Clinton but also willed into being by him. DeWitt was scion of New York's premier political family, the son of a celebrated federalist and first governor of New York State, George Clinton (1739–1812). DeWitt assumed his own presence in government, serving in the governments of both New York City and State. Clinton was a member of the New York State legislature (1798–1802 and 1806–1811), a U.S. Senator (1811–1815), and later served as Governor of New York (1817–1823). DeWitt Clinton had a compelling vision of the canal's success; Clinton is said to have assured the business community in New York City and members of the State legislature that the canal would make New York city "the greatest commercial emporium in the world" [5]. Three days after Clinton became Governor of New York State, construction on the canal began.

The Canal enabled both agricultural products and people to travel between New York City and the northern center of the country and provided a means for goods acquired through transatlantic or coastal commerce to reach the ports of the Great Lakes and the territories beyond the lakes.

Burrows and Wallace describe the immediate effects of the Erie Canal on towns and cities along its route [6]:

> Eight years after the first spade went into the ground and an amazing two years ahead of schedule, the great project was finally done – a marvel of human ingenuity and sacrifice by its engineers, who learned their trade on the job, and its laborers, many of them Irish and Welsh. Three hundred sixty-three miles long, forty feet wide, and four feet deep, the canal rose and descended a distance of 660 feet through eighty-three massive stone locks and passed over eighteen stately aqueducts....

> Within a year, the Erie boatmen were steering forty-two barges a day through Utica, bearing a thousand passengers, 221,000 barrels of flour, 435,000 gallons of whiskey, and 562,000 bushels of wheat. Shipping costs from Lake Erie to Manhattan plummeted from a hundred dollars a ton to under nine dollars. A few more years of this brought the annual value of freight transported along the canal up to fifteen million dollars, double that reaching New Orleans via the Mississippi; by mid-century the figure would approach two hundred million.

The Erie was the first but not the only canal that eased the flow of goods to New York; the success of the Erie inspired more canal-building. Before long, Burroughs and Wallace explain, a network of canals connected the agricultural and mineral producers of western New York, Pennsylvania, Ohio, Indiana, and Michigan to markets in New York.

As canals and rivers enabled agricultural and mineral products to flow to New York City, the Atlantic Ocean, a few miles East and South of New York harbor, provided a route for European immigrants to flood into New York. But it was not the ocean that enabled the "huddled masses yearning to breathe free"[4] to reach New York; it was the packet ships.

In its earliest use, the term "packet ship" referred to ocean-going vessels designed for scheduled transport of third-party goods between New York and Liverpool, England (Fig. 1.4). Soon thereafter, packet service to London and to Le Havre became available. The packets accepted shipments of various size and form; their scheduled departures enabled exporters to plan production schedules and importers on the other side of the Atlantic to have reasonable estimates of arrival times.[5]

---

[4] Language inscribed on the base of the Statue of Liberty is from the poem "The New Colossus" by Emma Lazarus (1883).

[5] The first packet service was announced in an advertisement that appeared in the New York *Evening Post*, October 27, 1817, under the title "Line of American Packets between N. York and Liverpool" Fig. 1.4. Following details of the first New York to Liverpool run, the copy advertisement continues: "It is intended that this establishment shall commence by the departure of the *JAMES MONROE*, from New-York on the 5th and the *COURIER* from Liverpool, on the 1st, of the First Month (January) next; and one of the vessels will sail at the same periods from each place in every succeeding month. ISSAC WRIGHT & SON, FRANCIS THOMPSON, BENJAMIN MARSHALL, JEREMIAH THOMPSON" [7]. This service instituted by these four individuals came to be known as the Black Ball Line.

Fig. 1.4 1817
Advertisement in *The New
York Evening Post* for packet
service between New York
and Liverpool [7]

LINE OF AMERICAN PACKETS
BETWEEN N. YORK & LIVERPOOL·

IN order to furnish frequent and regular conveyances for GOODS and PASSENGERS, the
subscribers have undertaken to establish a line
of vessels between NEW-YORK and LIVER-
POOL, to sail from each place on a certain day
in every month throughout the year.

The following vessels, each about four hundred
tons burthen, have been fitted out for this purpose:

> Ship AMITY, John Stanton, master,
> " COURIER, Wm. Bowne, "
> " PACIFIC, Jno. Williams, "
> " JAMES MONROE, —— "

And it is the intention of the owners that one
of these vessels shall sail from New-York on the
5th, and one from Liverpool on the 1st of every
month.

These ships have all been built in New-York,
of the best materials, and are coppered and copper fastened. They are known to be remarka-
bly fast sailers, and their accommodations for
passengers are uncommonly extensive and com-
modious. They are all nearly new except the
Pacific; she has been some years in the trade,
but has been recently thoroughly examined, and
is found to be perfectly sound in every respect.

The commanders of them are all men of great
experience and activity; and they will do all in
their power to render these Packets eligible conveyances for passengers. It is also thought, that
the regularity of their times of sailing, and the excellent condition in which they deliver their car-
goes, will make them very desirable opportunities for the conveyance of goods.

It is intended that this establishment shall commence by the departure of the JAMES MON-
ROE, from NEW-YORK on the 5th. and the
COURIER from LIVERPOOL on the 1st, of
First Month (January) next; and one of the vessels will sail at the same periods from each place
in every succeeding month,

> ISAAC WRIGHT & SON,
> FRANCIS THOMPSON,
> BENJAMIN MARSHALL,
> JEREMIAH THOMPSON.

10mo24

The idea for the packet ships is credited to Jeremiah Thompson, an Englishman
working in New York; the packet lines were imagined for transport of goods and not
particularly humans. The idea is this: an owner of a "line" of ocean-going vessels
would guarantee departure of a vessel on schedule. In January 1818, the first of the
Black Ball Line ships departed New York on time. The simple idea of scheduled
departures from New York appealed to businesses on both sides of the Atlantic. The
sellers of cotton in New York and the manufacturers of fabric in Manchester were
exactly such businesses.

Burrows and Wallace report that, as other packet line owners followed the Black
Ball Line's lead, "within two decades, fifty-two packets would be travelling regularly

from New York to Liverpool and Le Havre, an average of three sailings weekly, with an average transit time of thirty-nine days" [8].

The inventive piece in the institution of "packet lines" was their scheduled departures. The first packet line was announced in advertisements appearing in the shipping news in the New York newspapers. The announcement ran first on October 27, 1817 and explained that beginning the first weeks of 1818 four ships—the *Amity, Courier, Pacific,* and *James Monroe*—would carry passengers, mail, and freight on a monthly schedule between New York and Liverpool. Vessels would depart at specified dates and hour.

Historian Robert Greenhalgh Albion described the first packet sailing [9]:

At her East River pier lay the *James Monroe* seven months old and of 424 tons "burthen," the newest and, by a margin of forty tons, the largest of the four original packets. Little distinguished her from a dozen other Liverpool traders except the black ball at the top of her mainmast and the big black circle which would come into view when her foretopsail was unfurled. In her hold were fifteen hundred barrels of apples and smaller quantities of flour, cotton, and naval stores – a cargo of scant value compared with what the packets would later be carrying....

It was snowing hard that morning; that in itself might easily have been taken as an adequate excuse for delay. As St. Paul's clock struck ten, however, Captain James Watkinson gave the signal, the sails were trimmed, the lines were cast off and the *James Monroe* slid into the stream on time as advertised.

A great part of the cargo travelling westward was human. One could buy passage on a packet ship from Belfast or Liverpool to New York. As Burrows and Wallace explain [10]:

The result was that as more and more people left for the United States, more and more of them followed existing trade routes to New York. Between 1820 and 1832 the number of immigrants entering the port rose from thirty-eight hundred to some thirty thousand; in 1837 it swelled to nearly sixty thousand – almost 75% of the national total. Fed by this stream of humanity as well as internal migration, Manhattan's population climbed from 124,000 in 1820 to 166,000 in 1825 and 197,000 in 1830. By 1835 it exceeded 270,000. No other place in the country was growing so fast.

The passage between New York and Liverpool was one of four well-travelled sea routes into and from New York harbor. A second route to and from the piers at the shores of Manhattan followed Manhattan's eastern shore northward before veering eastward continuing through Hell Gate and the Long Island Sound to destinations on the country's northeastern shore. A third route led north on the Hudson River (which New Yorkers called the North River) to Albany and, with completion of the Erie Canal, west to the Great Lakes. The fourth route from New York harbor was to ports in the southern United States and in the Caribbean islands.

A ship exiting New York harbor could continue southward, following the coast of New Jersey, and further south to the Caribbean or to the "cotton ports" (Savannah, Charleston, Mobile, and New Orleans) in the southern United States.

There was strong cause for linking the cotton ports in the U.S. South to New York harbor: the making of clothing was a vigorous industry in New York, and New York had the county's best infrastructure and expertise for export. England and

France sought American cotton, and, as Albion explains [11], New York had the best facilities for providing it.

New York became the country's leading exporter of what the Register of the Treasury characterized as "colored" and "uncolored" manufactured American cotton fabric [12]. The Gulf Stream, the river of water in the Atlantic Ocean that flows North from the Gulf of Mexico, following the eastern coastline of the U.S. to the vicinity of New York then veers in a northeasterly direction, seemed to be in service of the New York shippers and their positions in the business of receiving American cotton from the South and exporting that cotton, either unfinished or in form of cloth, to England and Europe.

Concerning the Gulf Stream, Albion writes [9]:

> By a strange coincidence, it happened that the two major sea lanes which New York sought to appropriate for itself ran roughly parallel to that most remarkable of ocean currents, the Gulf Stream.... This lane of warm water results from the easterly trade winds which blow toward the Americas in the vicinity of the equator. These drive a vast quantity of warm water into the Gulf of Mexico and the adjacent Caribbean region and some of this finds an outlet into the narrow straight between Florida and the Bahamas. That is the beginning of the current which follows the American coast northward for some distance and then, off New York, bends eastward and crosses the North Atlantic to the British Isles in a somewhat diluted form. During the packet period, the whole current, from Florida to Ireland, was called the Gulf Stream.

New York's infrastructure for export was unmatched. When the chemicals manufacturing and the petroleum refining industries established themselves in New York harbor, they would make good use of the city's unrivaled ability to connect with foreign markets.

Another driver of nineteenth century New York's growing sophistication was the city's expanding intellectual life. Even before the United States existed, New York Island was home to an institution of higher education[6] and as various professional societies were forming in American cities in the late eighteenth and early nineteenth centuries, a few took form in New York.

New ideas, including those that concern applied chemistry, were amplified by presentations at scientific lectures and regularly reported in the New York publication "Scientific American." Book publishers in New York kept abreast of European scientific titles; science-focused manuscripts found publishers in New York. Chemistry became a subject of study at universities in New York City in this same period. The University of the City of New-York (later renamed New York University) was established in 1831 and began instruction in Chemistry the following year [14]. The Legislature of the State of New York chartered establishment of the Free University of New York in 1847. The Free University was later expanded and renamed The City University of New York [15]. Columbia University, the oldest of the New York universities, established a department of Chemistry in 1864.

---

[6] From the Columbia University website [13]: Columbia University was founded as "The College of the Province of New York, the City of New York," to be known as King's College, in 1754 by royal charter of King George II of England. It is the oldest institution of higher learning in the state of New York and the fifth oldest in the United States.

The newer universities, New York University and the City University of New York, were chartered with the explicit purpose of making higher education available to the whole population of the rapidly growing city: Albert Gallatin (1761–1849), one of the founders of NYU, described it "a social investment and a direct response to the needs of the rising mercantile classes in New York." The charter of Free University (which became the City University of New York) charges it "to offer a free, quality education based on academic worth, and to serve all social classes of the city" [15]. Indeed, providing higher education that is inclusive and accessible to all New Yorkers remains an ideal of several higher education institutions in New York. In the late 18th and early nineteenth centuries institutions of higher education were established in several cities along the nation's eastern seaboard, including Boston, Providence, New Haven, Philadelphia, and Baltimore, but those in New York were in the best location for influencing and receiving benefit from the City's sophisticated manufacturing economy, its financial markets, and its international trade.

The rapid growth in New York City's population, the new programs in higher education, the commercial networks that enabled goods from far-away parts of the country to flow to New York, and the scheduled seaborne trade with western Europe: all these advances in commerce and technology were initiated just as industrial production at Newtown Creek was advancing from tissue-recovery to fabrication of new materials. A rapidly expanding population in New York, providing both workforce and consumers of manufactured materials, and ready access to world markets, enabled by the vigorous harbor, provide two of three pieces needed to launch and support a vigorous materials manufacturing economy. The remaining piece was invention. The workforce and the consumers needed inventors, thinkers who understood production technologies and nurtured ideas of what might work better. How invention found application in chemistry and chemical processes over the course of the nineteenth and early twentieth centuries in New York City is the subject of this book.

As New York City secured its infrastructure for receiving raw cotton from the American South, sugar from the Caribbean islands, minerals and agricultural products from the western regions of the country, and as the City received thousands of new immigrants, a small group of New York business leaders organized the New York Chemical Manufacturing Company, the city's earliest enterprise with that specific focus.

Traditional medicines and dyes constituted much of the trade of the Chemical Manufacturing Company, but reagent chemicals, chemical compositions designed to have effect on other materials (sulfuric acid, "oil of vitriol," is the best example), were prepared as well. Cotton fabric would be a likely application for dyes, and reagent chemicals would be used in the salvaging of value from natural sources, dyes from trees and plants and crystalline sucrose from sugarcane. More complex materials could be had from the tissues of animals. Just as whales were slaughtered for their spermaceti and baleen, and pigs for their meat and hides, horses, the beasts of burden of nineteenth century American cities, were, immediately on their deaths, dissected for their valued tissues, hides, hooves, and bones.

In the middle decades of the nineteenth century, in the American city with the largest population and most sophisticated of infrastructures for trade and finance, an entrepreneurial culture and a materials-manufacturing economy began to grow.

Beginning in the early 1830s, the slaughterhouse industries that recovered value from the carcasses of animals clustered in corners of the harbor that were far from crowded city-center. Dead horses are malodorous, and the processes of tissue recovery are grisly. A rough business, the making of glue from hides and hooves of horses, and a second, the dissolution of the bones of horses in sulfuric acid solutions, grew and succeeded at the eastern end of Newtown Creek, a few miles east of Manhattan, far enough from the noses of the citizens of New York.

Industrial manufacturing at Newtown Creek, which began with the businesses of recovering value from the tissues of horses, consumed most of the Newtown Creek waterfront for more than a century. It is unlikely that Peter Cooper, on launching his new glue-making factory at Bushwick in 1830, contemplated filling the holds of packet ships with glue bound for Liverpool, but two decades later, when Abraham Gesner built a kerosene factory at Blissville on the northern shore of Newtown Creek in the early 1850s, he may have contemplated the volume of whale oil exported through New York harbor, and later in 1870, when the Standard Oil Company made its first acquisition, an oil refinery along the East River near Newtown Creek, it did so, one can believe, with export markets in mind.

Coincident with New York City's expanding network of commerce and trade in 1830s and 1840s was a new practice in the application of chemical science to agriculture. The manufacture of "artificial manure" (fertilizers produced in laboratories or factories) was developed, mainly in Germany and England, but the manufacturers of fertilizers in the United States and in New York stayed abreast of European teachings. A new chemicals industry was establishing itself in and around New York harbor.

This book concerns the growth of industrial chemistry in New York harbor, specifically at Newtown Creek, near the geographic center of modern-day New York City. That a chemicals industry grew in the harbor that was becoming the world's busiest harbor was not lost on the producers at Newtown Creek. It was this New York, the city to which agricultural and mineral goods from the bountiful states to the West flowed and into which risk-taking immigrants flooded from Europe, and at this time that materials-manufacturing industries took form along Newtown Creek. This narrative on industrial production begins at the eastern end of Newtown Creek, beyond Whale Creek, not far from the Dead Horse dock. This story begins at the abattoirs, the slaughterhouses that salvaged value from the hides and hooves of the city's dead horses.

This Newtown Creek story follows the succession of production technologies and the sustained growth in production scale and technological sophistication. Identifying the origin of the process is more difficult. If the "chemicals, fuels, metals, and New York City" story is about one technology supplanting another, one wonders, "What happened first? What triggered Newtown Creek's and the country's industrial expansion?" There are two answers to both questions: the chemistry answer is "a reliable supply of sulfuric acid;" the commercial product answer is "fertilizer." We begin with sulfuric acid.

References

1. Anderson M (2015). https://blog.epa.gov/2015/07/14/ the-manhattan-skyline-why-are-there-no-tall-skyscrapers-between-midtown-and-downtown. Accessed Aug. 3, 2021.
2. Lionel Pincus and Princess Firyal Map Division, The New York Public Library (1818) United States of America. https://digitalcollections.nypl.org/items/510d47da-eec7-a3d9-e040-e00a18 064a99.
3. Lionel Pincus and Princess Firyal Map Division, The New York Public Library (1820) Map of the Hudson ... from New York harbor the Fort Washington. https://digitalcollections.nypl.org/items/510d47db-c5d5-a3d9-e040-e00a18064a99
4. Lionel Pincus and Princess Firyal Map Division, The New York Public Library. (1820) Map of part of New York State between Albany and Buffalo: showing Erie Canal and other transportation routes. https://digitalcollections.nypl.org/items/510d47da-f056-a3d9-e040-e00 a18064a99.
5. Burrows EG, Wallace M (1999) Gotham: a history of New York City to 1898. Oxford University Press, New York, p 429
6. Burrows EG, Wallace M (1999) Gotham: a history of New York City to 1898. Oxford University Press, New York, pp 430–431
7. New York Evening Post, October 27, 1817. Digital image prepared by author at New York Public Library
8. Burrows EG, Wallace M (1999) Gotham: a history of New York City to 1898. Oxford University Press, New York, p 433
9. Albion RG (1938) Square-riggers on schedule; the New York sailing packets to England, France, and the cotton ports. Princeton University Press, Princeton, pp 21–22
10. Burrows EG, Wallace M (1999) Gotham: a history of New York City to 1898. Oxford University Press, New York, p 434
11. Albion RG (1939) The rise of New York port, 1815–1860. Charles Scribner's Sons, New York
12. United States (1877) Register of the Treasury. Commerce and navigation of the United States, Washington
13. Columbia University, History. http://www.columbia.edu/content/history. Accessed February 5, 2022.
14. New York University, A Brief History of New York University. https://www.nyu.edu/faculty/governance-policies-and-procedures/faculty-handbook/the-university/history-and-traditions-of-new-york-university/a-brief-history-of-new-york-university.html. Accessed February 5, 2022.
15. The City University of New York, History, Origins and Formative Years. https://www.cuny.edu/about/history/origins-and-formative-years/. Accessed February 5, 2022.

Chapter 2
Skin and Bones

In death the horse became a commodity as well. When horses became sick, lame, or just too old to justify the money spent in feeding and stabling them, their owners had them shot. It is likely that relatively few animals died directly from natural causes, but some did drop dead on city streets or in the stable....

As usual, New York had the most massive problem. Seven to eight thousand horses a year died in Manhattan in the early 1880s, compared with only fifteen hundred a year in Chicago. As many as thirty-six died on Manhattan streets daily during the Great Epizootic in 1872.

Clay McShane and Joel A. Tarr, *The Horse in the City, Living Machines in the Nineteenth Century,* The Johns Hopkins University Press, Baltimore, 2007, pages 27–28.

New Amsterdam began as a settlement at the southern tip of a long narrow rock-solid island. A deep harbor surrounds the island; along its western shore, a wide river flows from the North and offers a navigable route to places that a European explorer would know to be a new world. A broad channel to the south of New Amsterdam continues through two large bays and provides a navigable route to the Atlantic Ocean and passage to the old world from which the Netherlanders had sailed.

The city founded by Dutch trading ventures in the seventeenth century (as imagined in Fig. 2.1 in the nineteenth century) was constructed on land that was home to the Lenape and other native peoples. The success of the trading venture required workers; the Dutch venture welcomed settlers, including those English colonists who had grown weary of the orthodoxies of the colonies in New England. In time, English trading companies succeeded the Dutch trading companies that had established New Amsterdam; the established colony was renamed New York. Steadily, over the course of additional centuries, the city grew in population and prosperity. New York City has thrived for nearly four centuries.[1]

As the City of New York grew in population, the New Yorkers made their island grow. Land in Manhattan was reclaimed and reengineered to meet the needs of the city's burgeoning population. For centuries, successive populations have reimagined

---

[1] On this subject I am indebted to Russel Shorto's narrative on Manhattan [2].

© The Author(s), under exclusive license to Springer Nature Switzerland AG 2022
P. Spellane, *Chemical and Petroleum Industries at Newtown Creek,*
History of Chemistry, https://doi.org/10.1007/978-3-031-09629-7_2

Fig. 2.1 New Amsterdam a small city on Manhattan Island (Courtesy of The New York Public Library [1])

land in Manhattan: wilderness transformed to farmland, farmland to manufacturing sites, and manufacturing sites to land for new homes. Along the island's southeastern and southwestern shores, water lots were land-filled: dry land constructed from what had been piers in the harbor added mass and area to the precious island [3, 4]. In the early decades of the nineteenth century, New Yorkers demanded of the city space for homes, places to in which to live and raise families; New Yorkers then sought what a later generation would call "housing." As a city thrives, its use of land is continually reimagined.

In the first half of the nineteenth century, New York City grew to the north. Land to the south had already been built upon. New homes could be had to the East (across the East River, on Long Island) and to the West (in New Jersey across the Hudson River), but demand for homes in Manhattan continued. Then as now, many New Yorkers preferred to live in Manhattan.

Land near Kipps Bay, along Manhattan's eastern shore, a few miles North of the early settlement, was part of New York City, but not chic urbane New York. The Kipps Bay area was home to animal slaughterhouses and to the trades that recovered value from the tissues of freshly killed animals. In the 1830s and 40s, the Kipps Bay neighborhood underwent a process akin to what is now called "gentrification." The area of gritty slaughterhouses was transforming itself into a community of storefront businesses and family homes. In the present century "Kipps Bay" is not a bay; it is a neighborhood of storefronts and family homes at and around Third Avenue and East 30th Street.

The story of chemicals manufacturing at Newtown Creek began with the displacement of a glue factory. In the late 1830s, Peter Cooper relocated his glue factory from Kipps Bay to a site at the eastern end of Newtown Creek. Cooper's move followed the migration of the slaughterhouses and animal-tissue-salvaging operations from Manhattan to a site far from the sensibilities of the city folk settling into new houses at Kipps Bay. Peter Cooper and his glue-making business followed his suppliers of the hides and hooves.

Uprooted from its first venue, Cooper's production machinery could have been transported by harbor barges from the East Side of Manhattan to the eastern end of Newtown Creek. Newtown Creek's eastern end had become a dead-horse district, a place far enough from Manhattan and the communities at Greenpoint and Williamsburg that the gritty work of salvaging animal tissues could be carried out without disturbing the citizens of New York or Williamsburg or Greenpoint. In an area the Dutch had named Bushwick, Cooper built a factory near the trades that provided the hides and bones from which he extracted the collagen that forms glue. In building his glue factory at Bushwick, Cooper built at Newtown Creek a modern materials-manufacturing venue. Cooper's glue factory applied modern industrial technique to an ancient practice and launched at Newtown Creek a center for technology-driven manufacture of refined materials.

Cooper's biographer Edward Mack explained[2] [5]:

> By 1838, 33rd street had ceased to be a tenable place for a glue factory, so Peter moved it across the river to Maspeth Avenue in Brooklyn .... On May 16, 1838, he bought five acres to the north of the 'Maspeth Turnpike Road' in the then town of Bushwick; in 1841 he added four adjacent acres bordering on Wood Point Road, and subsequently eight more to make a tract running all the way to Newtown Creek.

Peter Cooper had a long and eventful life. He was born poor and died rich. He succeeded at business and in civic life repeatedly. As his life evolved, his achievements in business, wealth-making, and public life accumulated, and his generosity grew. Cooper became a celebrated presence in New York. Especially in the later decades of his long life, Cooper wrote extensively and participated in the cultural and political life of the city and of the nation. Appreciation among New Yorkers for Peter Cooper remains strong more than a century after his death.

Peter Cooper was born at the time of President Washington's inauguration. His schooling was minimal. An autodidact (not unlike several other industrialists who would follow Cooper in manufacturing at Newtown Creek), Cooper was, at age 17, independent, on his own, and determined to succeed. He began his career as an assistant to a machinist, a maker of devices. He took a second apprenticeship with a carriage-maker. Cooper had a rare combination of skills: he had an innate sense of mechanics and possessed what experimentalists refer to as "good hands." Even at the start of his career, he had the sense and confidence to make notes of his ideas

---

[2] Mack refers (his reference 29) to Index of Conveyances (for Brooklyn) book 76, p. 273; book 96 m p., 120; book 128, p. 393.

and drawings of his designs. Imagination was one of Cooper's gifts; skillful model making was another. He would build prototypes of the machines he envisioned. Early in his career, Cooper designed and constructed fittings for carriage axles; on another project, he improved the design of machinery used for shearing cloth.

As Cooper matured, his inventions grew more sophisticated and larger in scale. In Baltimore, Cooper built the Canton Iron Works, a steel-making plant, to provide rails for the Baltimore and Ohio Railroad. He designed steam engines and ferry boats and locomotives. He built the nimble Tom Thumb, a coal-fired steam locomotive that could haul trains through tight turns. Throughout his career, Cooper liked property, tangible and intangible, from land and buildings to patent rights.

In youth Cooper had no opportunity to study at university, but in age and maturity, he became a most fervent advocate for advanced education. Cooper chartered and endowed an extraordinary engineering institution in Greenwich Village. He supported the Cooper Union with funds adequate to provide free higher education to men and women. Cooper's endowment continues to defray the costs of students studying engineering at Cooper Union.

Cooper liked commerce, the give and take, the buy and sell of things of value, from groceries to land and buildings. He engaged in political discourse on matters of national concern, from abolition and the rights of native people to paper money and taxes. He hosted Abraham Lincoln in New York before Lincoln ran for national office.[3] After President Lincoln's assassination, Cooper advised Lincoln's successor, President Andrew Johnson. For most of his long career, even as he pursued various private and public ventures, Cooper manufactured glue (Fig. 2.2) and gelatin desserts.

Cooper's involvement in glue-making ((and Jell-O-making) began by chance. In the fall of 1821, Cooper owned a grocery shop to the north of New York City. His acquiring a glue factory seems the result of a neighbor's need to sell a business and Cooper's attraction to a good price and his inclination for new challenges. John Vreeland, the owner of the distressed glue-making factory not far from Cooper's grocery store, sought a buyer. Many years later, Cooper recounted the day [6]:

> I, hearing this, when he came by my house I stopped him and asked if it was true that he had a factory that he wanted to sell. He said he had a desire very much to sell it. I just stepped into my store, put on my hat and went up with him to look at it, and when I saw what he had there for sale and the price he asked for it I concluded at once to take it and told him that I would take it at the price at which he offered it. I went right downtown with him, without going home, and closed the bargain and paid him of it and went into a new business, from the grocery to the glue business, believing that if other people could carry it on I could try to do so. The situation of this was on the old Middle Road, occupying about three acres of ground which bounded it from Thirty-first to Thirty-fourth Street.

Cooper's biographer Edward Mack reports that Vreeland sold Cooper the lease to the 2.12 acres, the buildings, and all materials, stock in trade, fixtures, utensils, good and chattels on the premises except oil and bone on hand, and the lease of the land. The price was $2200; Cooper paid one thousand dollars on agreeing the price

---

[3] Lincoln addressed the Young Men's Republican Union at Cooper Union on February 27, 1860. Lincoln argued against allowing slavery in the nation's new western territories.

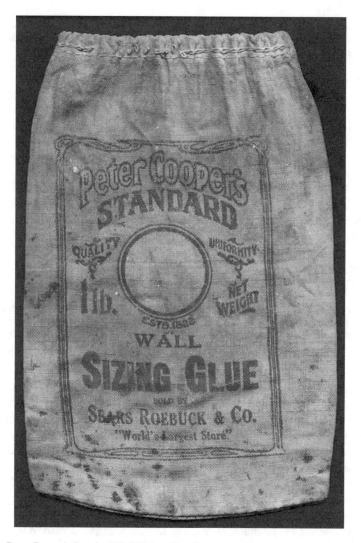

Fig. 2.2 Peter Cooper's Standard Wall Sizing Glue (image provided by Cooper Union Library [7])

and paid five hundred dollars plus interest May 1, 1822 and seven hundred dollars plus interest on May 1, 1823.

The planeness of glue disguises its profitability. Mack examined the business results of Cooper's glue-making business [5]:

> Before he moved out of Manhattan Peter was doing a large business in Philadelphia, Baltimore, New Haven, Albany, and Boston. And he had long since expanded his business to include not only glue but gelatin, isinglass, neat's foot oil, prepared chalk, and whiting (pulverized chalk cleansed of impurities); later he also ground white lead and did the milling for the manufacture of buckskin leather. In 1833 the stock in his factor was valued at $15,000, and the factory land lease of the ground at $10,000. Cooper himself admits that he 'had made

a good deal of money' by 1828. His grandson, who says that the glue factory 'was much more profitable than anyone knew.' Estimates that he made $10,000 to $20,000 a year from it while it was in New York.

When Cooper moved the glue factory to Bushwick, he did so with know-how earned in his 17 years in the trade. We must believe that Cooper, who made patented improvements to carriage-fittings and cloth shearing machinery, would have engineered glue-making machinery and methods to his own design. Shortly before the move to Bushwick, Cooper was awarded US Patent 5943x, published April 29, 1830 (Fig. 2.3) [8], which taught the importance of temperature control in condensing glue stock, the watery soup of collagen (protein) extracted from the hides and hooves of animals. His patent describes how a farm-kitchen technology can be improved by attention to technique: glue-making optimized by process control. "Hide glue" consists of proteins extracted from animal tissues into warm alkaline water. One would prepare glue-stock in the manner that a chef prepares chicken-stock, the watery solution of collagen prepared by slow-simmering the bones of chickens.

The three-dimensional shapes of proteins are sensitive to temperature. Professor Francis Carey of the University of Virginia explains [9]:

> The way a protein chain is folded, its tertiary structure, affects both its physical properties and its biological function. The two main categories of protein tertiary structure are fibrous and globular. 1. Fibrous proteins are bundles of elongated filaments of protein chains and are insoluble in water. 2. Globular proteins are approximately spherical and are either soluble or form colloidal dispersions in water.

Just as the runniness or rubbery-ness of the whites of soft-boiled eggs depends on the time and temperature of their boiling, the proteins Cooper extracted from the tissues of horses were vulnerable to denaturation, an unfolding of their native form, at elevated temperatures. Cooper recognized that the process of recovering the (protein) material he extracted from the bones, hides, and hooves of horses can determine the quality of the product. Along with new machinery for the manufacture of glue, Cooper brought to Newtown Creek technological sophistication. He raised the tissue-salvaging area's technology game as he introduced a new and patented process control; he understood that control of the conditions of a process, variables like time and temperature, matter. The Cooper factory recognized the role that chemical technology can play in determining product quality.

Within months of Peter Cooper's move to Bushwick near the end of the 1830s, an ambitious businessman and experienced chemist, Martin Kalbfleisch, established his third and most successful chemicals manufacturing operation at a site that adjoined land owned by Cooper. The area's unspoken competition, its practice of matching each innovative technology with a cleverer technology, its game of one-up-man-ship, was on.

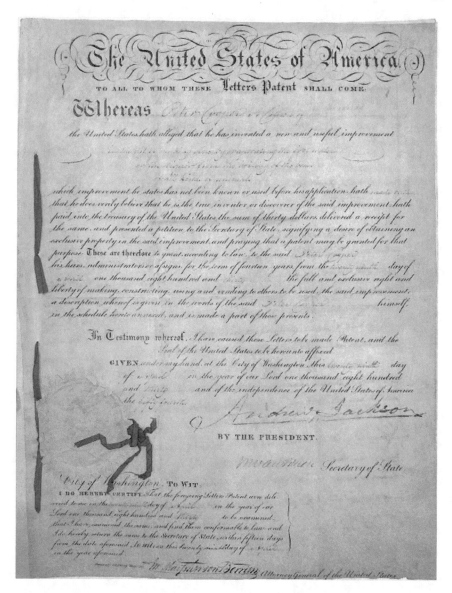

Fig. 2.3 US Patent awarded to Peter Cooper [8]. Note signatures by President Andrew Jackson and Secretary of State Martin van Buren (Image provided by Cooper Union Library)

References

1. The Miriam and Ira D. Wallach Division of Art, Prints and Photographs: Picture Collection, The New York Public Library (1857) *New Amsterdam a small city on Manhattan Island.* https://digitalcollections.nypl.org/items/510d47e1-2ba2-a3d9-e040-e00a18064a99
2. Shorto R (2004) Island at the center of the world. Vintage Books, New York
3. Schlichting K (2018) Manhattan waterfront. Johns Hopkins University Press, Baltimore
4. Burrows E, Wallace M (1998) Gotham: a history of New York City to 1898. Oxford University Press, New York
5. Mack E (1949) Peter Cooper, citizen of New York. Duell, Solan & Pearce, Inc. New York, p 194
6. The Autobiography of Peter Cooper, 1791–1883, dictated by him February 20 to April 17, 1882. Available at Cooper Union Library, "Transcribed from the original shorthand notes at Jersey City, N. J., April 1948. By: William S. Coloe, Certified Shorthand Reporter of New Jersey. This excerpt is from page 180
7. The Cooper Union Library, The Cooper Union Archives & Special Collections. https://library.cooper.edu/archives. Accessed February 5, 2022
8. US Patent 5943x, published April 29, 1830. Image provided by Cooper Union Library
9. Carey F (2006) Organic chemistry. McGraw Hill Higher Education, New York

Chapter 3
Oil of Vitriol: Martin Kalbfleisch and the Manufacture of Reagent Chemicals at Newtown Creek

We may fairly judge of the commercial prosperity of a country from the amount of sulphuric acid it consumes.

Justus von Liebig, Familiar Lectures on Chemistry (1843)

An early map of the place we now call Brooklyn identifies itself "Map of Kings and part of Queens counties, Long Island N.Y./ surveyed by R.F.O. Conner (1852)" (Fig. 3.1) [1]. The map provides extraordinary detail: names of roads and streets, sites of churches and graveyards, locations of about 150 houses in Greenpoint, routes of ferries that crossed the East River, and details of the U.S. Navy Yard, including the site and footprint of the U.S. Naval Hospital that still stands there. In less settled parts of the county, Conner's map provides the names of owners of large tracks of land. Not many industrial sites are noted, but at the south and eastern end of Newtown Creek, adjoining land identified with the names P. Cooper or W. Cooper, a site that that includes several buildings is labelled "M. Kalbfleisch Chemical Works."

In the early 1840s, Martin Kalbfleisch established a new production plant at Bushwick, a village that adjoins Newtown Creek at its southeastern end. Kalbfleisch was, at that time, an established and experienced chemist and business executive. He had built and sold manufacturing businesses in Harlem (a village in northern Manhattan) and in Bridgeport, Connecticut. After a short tenure in Bridgeport, Kalbfleisch returned to the New York area to build and launch another new business. One can assume that this accomplished professional would have good cause to abandon his Bridgeport venture and return to New York. Kalbfleisch did not return to Manhattan, where he had begun his career in chemistry; he chose to live in Greenpoint, a small city across the East River from Manhattan. Greenpoint is a part of New York harbor, with waterfront on both the East River and Newtown Creek. Kalbfleisch made a home in Greenpoint as he built a new manufacturing plant a few miles east at Bushwick.

P. Spellane, *Chemical and Petroleum Industries at Newtown Creek*, History of Chemistry, https://doi.org/10.1007/978-3-031-09629-7_3

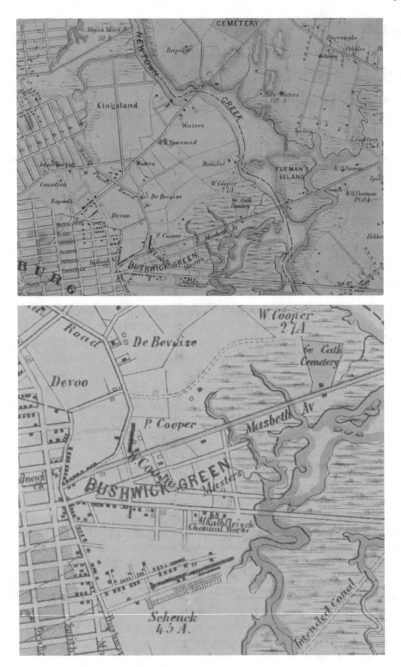

Fig. 3.1 Details of "Map of Kings and Part of Queens Counties, Long Island, NY" (Courtesy of The New York Public Library [1])

Martin Kalbfleisch had begun life in the Netherlands in 1804[1] [2, 3]. Records suggest that he grew up in fortunate circumstances: the child of a prosperous burgher and beneficiary of a good education in Vleissingen. His schooling may have provided Martin an introduction to chemistry. Despite the apparent comfort of life in Vleissingen, Kalbfleisch's ambition and curiosity overwhelmed any filial and scholarly attachments that he may have had for home. At age 18, Kalbfleisch set out on a career as a merchant seaman, sailing on an American ship (the three-masted windjammer "Ellen Douglass," homeport Salem, Massachusetts) bound for the islands that Kalbfleisch would have known as the Dutch East Indies.

The voyage was ill-fated. Cholera on the island of Sumatra prevented the crew from disembarking and forced the Ellen Douglas to return to Europe. Having sailed from western Europe to the East Indies and back, Kalbfleisch sailed on to Paris where he joined his American skipper in a business venture. With the experience of distant travel and the ineffable benefits of time spent in Paris as a young man (in Martin Kalbfleisch's case, those included an English wife, a son, and opportunity to hear lectures on chemistry at the Sorbonne), the 22-year-old Kalbfleisch and family embarked for New York. Soon after arriving in New York, the new New Yorker found work in the city's nascent chemicals industry.

Martin Kalbfleisch arrived in New York in 1826. In 1826, the industrial production of chemicals in the United States had barely begun. Philadelphia was the young country's leading center for manufacture of chemicals. John Harrison, who had apprenticed with the celebrated chemist Joseph Priestley,[2] set up the country's first facility for the industrial production of sulfuric acid by the then state-of-the-art "chamber process" in Philadelphia in 1793. The Philadelphia partnership Harrison and Lennig remained the country's premier producers of sulfuric acid for years, adapting best methods and practices as they became known (e.g., replacing glass retorts with platinum stills to further concentrate the chamber acid). New York's chemicals industry was a distant second to Philadelphia's.

Sulfuric acid is easy to make but difficult to concentrate, difficult to transport, and notoriously corrosive, yet it was then (and is now) essential to every important process in industrial chemistry. If New York held ambition of becoming a modern and commercially viable city (that is, having reliable sources of dyes and paints, medicines, bleach, and soap), the city's economy would require, at minimum, one reliable supplier of sulfuric acid.

Thinking of that sort inspired a group of accomplished business leaders in New York to launch a chemicals venture. They selected an ingenuous name and articulated, in their founding documents, a clear mission. The New-York Chemical Manufacturing Company was chartered in 1823. The company's name was more aspirational than descriptive of its founders' business interests.

---

[1] Biographical information on Martin Kalbfleisch is found in a work published during Kalbfleisch's lifetime by Henry R. Stiles [2]. At the time of that book's publication, Kalbfleisch was serving his second term as Mayor of Brooklyn.

[2] Priestley, an Englishman, had met Benjamin Franklin in London in 1776. Priestley's scientific work was extensive and well received, but his opinions on religion were not. Priestley was eventually forced to flee England. Franklin helped Priestley settle in the United States in 1794.

In standard practice, corporations that were established in New York County were chartered at the office of the New York County Clerk. The new chemicals-manufacturing venture was not chartered there; the company's founders, John C. Morrison, James Jenkins, Gerardus Post, and Charles G. Haynes sought charter by the Legislature of New York State. "The New-York Chemical Manufacturing Company" was created by an act of the New York State legislature in 1823 [4]. The company constructed a manufacturing plant in Greenwich Village, in Manhattan, a few miles north and west of the city. The company's production site is pictured in Fig. 3.2.

The company's chartering document includes this language:

> BE it enacted by the People of the State of New-York represented in Senate and Assembly, That John C. Morrison, and such others as now are, or hereafter may be associated him for the purpose of carrying on the business of manufacturing blue vitriol, alum, oil of vitrial (sic), aqua fortis, nitre acid, muriatic acid, alcohol, tartar emetic, refined champhor, salt-petre, borax, copperas, corrosive sublimate, calomel and all other drugs and medicines, paints, and dyers articles, in the city and country of New-York, shall be, and hereby be ordained, constituted and appointed to be a body politic and corporate, in fact and in name, by the name of the "New-York Chemical Manufacturing Company"…

> XI. *And be it further enacted,* That the corporation hereby erected shall not engage in any banking business or transaction whasotever (sic), or in any other business or transaction,

Fig. 3.2 New York Chemical Manufacturing Company (Courtesy of the JPMorgan Chase Corporate History Collection)

excepting such as may be proper and necessary to carry into effect the declared objects of this act.

XII. *And be it further enacted,* That it shall and may be competent for the legislature at any time or times hereafter to alter, modify, or repeal this act.

XIII. *And be it further enacted,* That this act shall continue for the term of twenty-one years."

The first paragraph provides specific information on the company's intended chemical products. Paragraph XI is surprising, and paragraph XII is revealing. The last three paragraphs would be invoked the following year and 20 years after that. The charter of the new manufacturing company states that this manufacturing company will not be a bank, but in the following year, the New York State Legislature amended the charter of the Manufacturing Company [5]. The new act, "An Act to amend an act entitled 'An Act to Incorporate the New York Chemical Manufacturing Company,' Passed April 1, 1824," specified that [5]:

> Gerardus Post, together with James Jenkins and John C. Morrison, named in the seventh section of the act hereby amended, be and they are hereby appointed commissioners, and are authorized to receive subscriptions to the capital stock of the said corporation, to the amount of five hundred thousand dollars, in shares of twenty-five dollars each, and that the said corporation shall have the power to employ the whole thereof, excepting the sum of one hundred thousand dollars, in banking operations in the city of New-York, and to issue bills, notes or obligations under the seal of the said corporation, or without their corporate seal, as other banks in this state are authorized to do, and in such manner as the said corporation shall direct, and to make all proper rules and regulations, and to appoint all proper officers, clerks and agents, for carrying on the same...

Chemicals manufacturing in New York City had an inauspicious beginning. Within a year of its grand launch and publication of its ambitious but uncomplicated mission, the company that appeared destined to become city's premier chemical manufacturer transformed itself into a bank with a small manufacturing division. At the end of the original company's 21-year charter, despite its success in the business of making and selling chemicals, the directors elected to end its manufacturing work and focus exclusively on banking. After liquidating its inventory and physical assets, the company prospered as a bank and matured to become the Chemical Bank of New York. In time, Chemical Bank merged with the Manufacturers Hanover Bank, and all, in further time, became part of JP Morgan Chase and Company.

But, early in the company's money-handling days, the Chemical Manufacturing Company continued to make and sell chemicals. At that time, with the stress of new banking operations, the New York Chemical Manufacturing Company would find itself in need of hands trained in chemistry. John Morrison, the company's principal chemist, whose experience in manufacturing chemicals was central to formation of the original Chemical Manufacturing Company, had taken on new money-handling responsibilities. When the ambitious Martin Kalbfleisch, newly arrived in New York, sought work in the chemicals industry, the Chemical Manufacturing Company created the title "assistant superintendent and assistant to" the company's merchant-owner-chemist-banker John Morrison.

Kalbfleisch thrived with the new work. Within a year, the company having moved operations from Morrison's original location in Greenwich Village to a site near present day 32nd Street, Kalbfleisch was placed in charge of operations at the Chemical Manufacturing Company. He held that position until 1829 at which time he launched a business of his own, a dye manufacturing company, at a location along the Harlem River at Manhattan Island's northern end. The dye company was the first of Kalbfleisch's several companies. As land values rose, Kalbfleisch sold the dye business and proceeded to build a new business in Bridgeport, Connecticut. That venture did not last long. In 1839 or 40, Kalbfleisch and family returned to the city of their arrival in America and made a new home in Greenpoint, Long Island, a ferryboat ride across the East River from Manhattan. Secure in his home in Greenpoint, Kalbfleisch would give serious thought to the venue and goals of his next venture.

He organized his next and boldest manufacturing venture. It is likely that Kalbfleisch would be aware and perhaps attentive to Peter Cooper's glue-making factory at Bushwick. Cooper and Kalbfleisch may have known one another: the sites of the original glue factory and New York Chemical Manufacturing Company's new factory were within walking distance of one another. Having opened and closed his second company in Bridgeport, Kalbfleisch would choose a venue and business focus carefully. In 1841, he established his newest chemicals-manufacturing venture at Bushwick Green near the intersection of Maspeth Avenue and the Williamsburg and Jamaica Turnpike and not far from Peter Cooper's new glue factory. Cooper was 15 years older than Kalbfleisch; each had an established business, and both were involved in civic matters. At their manufacturing sites at the eastern end of Newtown Creek, Cooper would be Kalbfleisch's nearest neighbor and likely customer.

Cooper's glue factory may have attracted Kalbfleisch to Bushwick, but there was a more compelling cause for Kalbfleisch to build a sulfuric acid factory among the slaughterhouses and tallow factories. A new industry, the manufacture of "artificial manure," was taking form. The significance of the word "fertilizer" was evolving, adapting to a new chemical technology. The term that had once described the animal product that one finds readily in barnyards and pigpens and horse stables was being applied to a manufactured chemical. The idea that fertilizer could be manufactured followed several observations of the dissolution of phosphate-rich bones or phosphate minerals in sulfuric acid; the idea is evident in literature and reports from the 1830s. The grander idea, that agriculture could be enhanced by industrial chemistry, was articulated most compellingly by Justus von Liebig in the early 1840s [6]. "Artificial manure" describes the phosphate-rich residue of animal bones that have been treated with or dissolved in sulfuric acid.

A second term, one that seems modern even in the twenty-first century, superphosphate, soon replaced artificial manure. Superphosphate, a factory-made nutrient for plants, launched a large and profitable new chemistry-driven industry [7]. Superphosphate is the chemical offspring of sulfuric acid, the king of chemicals, and the phosphate-rich bones of animals. In placing his sulfuric acid factory in the dead-animal zone around the English Kills at Bushwick, Kalbfleisch positioned his newest venture between the bones of slaughtered horses and the new superphosphate industry.

A contemporary historian of chemistry described the Kalbfleisch factory's capacity for production of sulfuric acid as having a lead position in the country [8]:

Martin Kalbfleisch and the "Bushwick Chemicals Works … are among the most important and extensive Chemical manufactories in the United States. … One of the chambers, for manufacturing Sulphuric Acid, is two hundred seventeen feet long by fifty feet wide, no doubt the largest in existence, and is a model in every particular. Among the noticeable objects that attract the attention of visitors, are three Platina Stills, imported from France, at a cost of about fifteen thousand dollars each. … Of Sulphuric Acid they have a capacity for producing three hundred thousand pounds weekly.

In 1880, Scientific American's "American Industries" series featured the Kalbfleisch business and identifies the capacity and multiple production sites of the company's factories [9]:

The works were originally started in 1829, with one small factory for the production of sulphuric acid, but there are now five factories for this branch of the business, besides those devoted to the other specialties, the buildings and yards covering about twenty acres of land on Newtown Creek, at a point which can be reached by vessels drawing nine feet of water, but yet within the city limits of Brooklyn, N. Y. The firm also have extensive works of a similar character at Bayonne, NJ and Buffalo, NY, with which localities they have thought it best to divide their business because of its rapid growth of a few years past.

References

1. Lionel Pincus and Princess Firyal Map Division, The New York Public Library. Map of Kings and part of Queens counties, Long Island N.Y. The New York Public Library Digital Collections. 1852. https://digitalcollections.nypl.org/items/03910ed0-cd02-0133-9016-00505686a51c
2. Stiles HR (1869) History of the City of Brooklyn. Published by subscription, Brooklyn, NY, pp 492–494
3. Anon (1935) The Honest Dutchman' and His Youngest Son. Chemical Industries XXXVII(1):18–22. Note that Williams Haynes, editor of Chemical Industries at that time, is the likely author of this piece. Haynes had received notes on Kalbfleisch prepared by August Klipstein Sr., a contemporary of Martin Kalbfleisch, from August Klipstein, Jr. (correspondence and papers are in the Williams Haynes archive at the Chemical Heritage Foundation). Klipstein Sr. had built a chemicals-importing company in New York, a few decades after Kalbfleisch established his company. The Klipstein company and the chemicals company of Franklin Kalbfleisch, the youngest son of Martin Kalbfleisch, became parts of the American Cyanamid & Chemical Corporation in New York in the years following World War I
4. New York (1777) Laws of the State of New York passed at the sessions of the Legislature. New York (State), Albany, N.Y., p 37. https://babel.hathitrust.org/cgi/pt?id=osu.32437123260537; skin=2021;sz=25;q1=chemical;start=1;sort=seq;page=search;seq=57;num=37
5. New York (1777) Laws of the State of New York passed at the sessions of the Legislature. New York (State), Albany, N.Y., pp 140–141
6. Liebig J (1841) Organic chemistry in its application to agriculture and physiology. J. Owen, J. Munroe and company, Cambridge

7. Gray A (1945) Phosphates and superphosphates. Interscience Publishers, New York
8. Bishop J (1868) A history of American manufactures from 1608 to 1860, vol 3. Edward Young & Co, Philadelphia, p 196
9. Anon (1880) American Industries - No. 60. The manufacture of chemicals. Sci Am XLIII 21:322–323

Chapter 4
Superphosphate

"Superphosphate" describes a fertilizer composition that is rich in water-soluble phosphate salts, prepared either from phosphate-rich minerals or from the phosphate-rich bones of animals. Using finely ground bones as fertilizer is traditional, but bone or mineral dissolved in acid, because of its solubility, earned its own name. The sense of the word "superphosphate" is uncomplicated, but the origin of the word is the stuff of folklore.

Unlike the Edison lightbulb and the Schottky diode, there is no family name attached to superphosphate. It was not so much invented as it was recognized and acknowledged. Its history suggests that superphosphate thrived as a shared idea that matured, over the course of several decades in the first half of the nineteenth century, into an item of commerce. Chemists and farmers shared ideas, theories, and findings of diverse experiments, creating a folklore of "luxuriant crops" grown in soil nurtured with "vitriolized bones." Gaining credibility through papers presented at scholarly meetings and patents awarded, superphosphate won advocates and became an item of commerce, the earliest "artificial manure".[1]

In *Phosphates and Superphosphates* [2], A. N. Gray ascribes the earliest allusion to the idea of superphosphate to an entry in *Nicholson's Dictionary* (1808). Gray reports that, under the heading "Phosphoric Acid," the following text appears [2]:

> Whole mountains in the province of Estremadura in Spain are composed of this combination of phosphoric acid and lime… and when mixed with sulphuric acid forms sulphate of lime with the sulphuric acid whilst the phosphoric acid is set at liberty in the fluid.

James Murray of Ireland presented a paper at Dublin in 1857 that included mention of a report by Mr. S. Ferguson in 1808 of the benefit to plants of vitriolized bones in soil. Murray told of an alternative source of soluble phosphate [3]:

> This mineral or rock contains phosphoric acid, in combination with lime; common strong acids take the lime, and set free the phosphoric acid, which is then ready to unite with any alkaline material, such as soda, potash, or ammonia.

---

[1] A review of the subject, including its early history, is provided in the report "Superphosphate: its history, chemistry, and manufacture," prepared by the U.S. Department of Agriculture [1].

© The Author(s), under exclusive license to Springer Nature Switzerland AG 2022
P. Spellane, *Chemical and Petroleum Industries at Newtown Creek*,
History of Chemistry, https://doi.org/10.1007/978-3-031-09629-7_4

Joseph Graham, owner of bone mills in London, had prepared a pamphlet "A Treatise on the Use and Value of Bone Manure" in 1839 [4] that contained the following advice:

> Every substance constituting the food of plants, must first be reduced into a state of solution, before it can in any way be absorbed by their roots. ... Although phosphate of lime (as it exists in bone) is totally insoluble in water, yet it is well known to all chemists that this salt, when deprived of a portion of the lime constituting its base, and reduced into the state of a super (or acid) phosphate, becomes soluble, and that in exact ratio of the lime extracted.

Beginning in 1831 and continuing for more than a decade, Heinrich Wilhelm Köhler manufactured a chemically treated bone fertilizer at his factory in Mies, near Pilsen, Bohemia. Köhler's Austrian patent, granted in June 1, 1831, read in part, "According to my method the bones are treated with sulphuric acid, thus liberating a certain quantity of phosphoric acid...."

In 1835, Gotthold Escher, the headmaster of a school in Brünn, Moravia, suggested acid treatment of bones [5]:

> It is not improbable that bone meal would have a more satisfactory and more rapid fertilizer action if it were decomposed in a shorter time than that taken by the factors previously mentioned: light, air, heat, moisture, etc. Thus an attempt slightly to moisten bone meal prior to its immediate application to the fields with a cheap and not too strong an acid may prove successful.

When chemistry was a province of science but not yet an industry, fertilizer was uncomplicated. Animal-keeping farmers, one expects, would have access to and familiarity with natural fertilizers and understand the benefit they provide to growing plants. Two developments in the nineteenth century, the examination of the role of chemicals in plant growth and the industrial production of chemicals, enabled the industrialization of farming. The industrial age applied chemical science to farming and made farming more complex and more productive.

Developments in England (experimentation, the systematic examination of farming practices) and in Germany (scholarship, a close examination of the chemistry of plant development and growth) preceded the chemical industry's campaign to sell chemical products to farmers. That inquiry and experimentation began, in the modern period and among the European communities, with the scholarship of Justus von Liebig in Germany and the experimentation of John Bennett Lawes in England. Liebig, publishing in both German and English, was an energetic advocate for the science of chemistry in general and particularly for articulation of a chemical description of plant growth. He lectured in Germany and England and published an exhaustive examination of the chemistry of plant growth and the role of fertilizers. Liebig's text, *Organic Chemistry in its Application to Agriculture and Physiology*, treated farming as science; it specified the nutrients essential to a growing plant and advised the reader that the chemicals industry could provide the essential reagents that growing plants need. The second edition of Liebig's text was published in the United States in 1841 [6]. Liebig argued in favor of "scientific principles" and "artificial manures." In his introduction to the work, he writes on page xix [6]:

Perfect agriculture is the true foundation of all trade and industry, -- it is the foundation of the riches of states. But a rational system of agriculture cannot be formed without the application of scientific principles; for such a system must be based on an exact acquaintance with the means of nutrition of vegetables, and with the influence of soils and action of manures upon them. This knowledge we must seek from chemistry, which teaches the mode of investigating the composition, and of studying the characters of the different substances from which plants derive their nourishment.

In England, John Bennett Lawes studied the chemistry of plant growth and constructed an experimental station, perhaps the world's first experimental agricultural science lab, at his estate in Rothamsted. With Joseph Henry Gilbert and the American chemist Evan Pugh (who would return to Pennsylvania to become first president of the Agricultural College of Pennsylvania in 1860), Lawes prepared "superphosphate."

Lawes, having patented in England methods for preparing superphosphate from mineral sources, proposed that dissolving phosphate-rich minerals in sulfuric acid provides bio-accessible phosphate-rich fertilizer, and he established a fertilizer business: in 1843, Lawes established a company, "J. B. Lawes Patent Manures" with a factory at Deptford Creek [7].

At nearly the same time, Liebig described a method for releasing phosphate ion from bones. Animal bones are rich in phosphate; bones are dissolved in sulfuric acid. In a report to the British Association in 1840, Liebig recommended breaking down animal bones with sulfuric or hydrochloric acid and supplying plants a nutrient that had been "digested prior to application." He wrote, "the most easy and practical mode of effecting their (bones) division is to pour over the bones, in a state of fine powder, half of their weight of sulphuric acid, diluted with three or four of water, and after they have been digested for some time to add 100 parts of water, and sprinkle this mixture over the field before the plough."

Liebig and Lawes's examinations of the chemistry of plant growth expanded the understanding of the process, new knowledge that might have expanded the vocabulary of practical agriculture, but it appears that that did not happen. The word "manure" was burdened with new and various meanings. A forty-page pamphlet published in 1857 by fertilizer manufacturers H. and T. Proctor provides a useful guide to the variety new meanings of the word "manure". From page 6 of the pamphlet [8]:

At one time it was believed that farm-yard manure was the only efficient fertilizer to accomplish this desirable end; the experience of the last thirty years, however, has shown that, in great measure, artificial manures may be employed with advantage instead of yard manure, and that in many respects artificial manures deserve the preference to ordinary dung. It is upon the judicious selection of artificials and their proper application to the land, that the future progress of agriculture mainly depends.

The authors' "artificial manures" include these:

Peruvian guano

Nitrate of soda

Bone dust

Superphosphate of lime

Special manures for turnips, grass, wheat, potatoes, &c, &c.

Their "special manures" include the following without detail of the composition of each:

Turnip manure

Mangold manure

Superphosphate of lime

Wheat manure, no. 1

Wheat manure, no. 2

Barley manure

Oat manure

Potato manure

Carrot manure

Vetch manure

Pea and bean manure

Prepared bone, no. 1

Prepared bone, no. 2

Prepared bone, no. 3

Clover manure, no. 1

Clover manure, no. 2.

Their "List of Manures" includes the following: bone (drill size), bone dust, guano (Peruvian),
German compost, superphosphate of lime, prepared bone acid, sulfate of ammonia, nitrate
of soda, gypsum, sulfuric acid, and "special manure for sugar."

Just as "manure" would grow to include the "artificial manures," the word "fertilizer" would be applied to various compositions of matter. In the second half of the nineteenth century, the state of New York was one of the country's leading producer of fertilizers. In the present examination of industrial chemistry at Newtown Creek, we find evidence of numerous fertilizer companies at sites along Newtown Creek; many adjoin chemicals manufacturing sites. In the absence of records of their operations, and because of their proximity to chemical businesses, we propose that the fertilizers produced at Newtown Creek were likely to be superphosphate.

Superphosphate married industrial chemistry to agriculture. Superphosphate could be manufactured at factories in industrial corridors of cities, far from the farms or plantations that would consume it. Superphosphate, a factory-made fertilizer, enabled New York to become a leading producer of a commodity that agriculture would soon deem essential.

Gray's *Phosphates and Superphosphates* was published in 1945; the text cites data from 1938: worldwide consumption of phosphate in 1938 was just under 12 million tons; superphosphate accounted for slightly less than 10 million tons. Arthur Toy and Edward Walsh provide more recent (1987) data [9]:

> In the 1940s, superphosphate was used in about 90% of the domestic phosphate fertilizers. This amount decreased to about 30% in the 1960s, and presently about 8% of the domestic phosphate fertilizers contain superphosphate, as the usage of ammonium phosphate in solid, liquid, and suspension form continues to increase.

In the late nineteenth century, New York State was among the largest producers of fertilizer in the United States; only four states in the southern US, Georgia, North Carolina, South Carolina, and Virginia, produced more. The United States produced and consumed more phosphate (a component of many fertilizers) than did any other country. The two results, New York's production of fertilizer and the nation's consumption of phosphate, point to the importance of an industrial commodity, superphosphate, in the nation's agricultural economy.

That New York became, during the nineteenth century, a leading producer of fertilizer supports the understanding that "fertilizer" had evolved from specifying various organic and inorganic compounds found in nature, or easily produced from natural products, to describing a manufactured product. Superphosphate demonstrates that a production chemical can command a significant place in the nation's agricultural economy; superphosphate opened a large market, agriculture, for other industrial chemicals.

John Jay Mapes, who had been born in Maspeth, not far from Newtown Creek, took interest in many endeavors including agriculture: he edited a monthly periodical "The Working Farmer" printed at an office in New York City at 351 Broadway. In articles published between 1849 and 1852, he advised farmers of the fertilizing advantages of superphosphate. In 1852, Mapes established his own superphosphate production company in greater New York harbor at Newark, New Jersey.

Evidence of early production of superphosphate at Newtown Creek is imperfect. The area was home to numerous "fertilizer" companies; the Newtown Creek economy included both bones, from animal slaughterhouses, and producers of sulfuric acid. It is unlikely that fertilizer produced at Newtown Creek could be anything other than superphosphate, but the word "superphosphate" appears on insurance maps of Newtown Creek only later in the nineteenth century.

References

1. Soil and Water Conservation Research Division, Agricultural Research Division (1964) Superphosphate: its history, chemistry, and manufacture. U.S. Department of Agriculture and the Office of Agricultural and Chemical Development, Tennessee Valley Authority
2. Gray A (1945) Phosphates and superphosphates. InterScience, New York
3. Murray J (1858) On the choice of perennial rather than annual fertilizers. Abstract in Notices and Abstracts, Brit Assoc Adv Sci Rpt, 27th Meeting, 1857: 54–55

4. Graham J (1839) A Treatise on the Use and Value of Bone Manure. James Ridgway, London
5. Escher G (1835) The Production of and Fertilisation with Bone Meal as Carried Out on the Strassnitz Estate. Mittheilungen der Brünner Ackerbaugesellschaft
6. Liebig J (1841) Organic chemistry in its application to agriculture and physiology. John Owen, Cambridge
7. Warington R (1900) Sir John Bennet Lawes, Bart., F.R.S. Nature 62:467–468
8. Proctor H, Proctor T (1857) Manures: their properties and application. William Hodgetts, Birmingham
9. Toy A, Walsh E (1987) Phosphorous chemistry in everyday living. American Chemical Society, Washington DC

Chapter 5
Abraham Gesner and The New York Kerosene Oil Company

In merchantmen, oil for the sailor is more scarce than the milk of queens. To dress in the dark, and eat in the dark, and stumble in darkness to his pallet, this is his usual lot. But the whaleman, as he seeks the food of light, so he lives in light. He makes his berth an Aladdin's lamp, and lays him down in it; so that in the pitchiest night the ship's black hull still houses an illumination.

See with what entire freedom the whaleman takes his handful of lamps—often but old bottles and vials, though—to the copper cooler at the try-works, and replenishes them there, as mugs of ale at a vat.

From Chapter XCVII, The Lamp, Moby-Dick, or the Whale by Herman Melville, published by Harper and Brothers in New York, November 1851.

Moby-Dick appeared in 1851, by which time, whale oil had become too costly to burn. As a fuel for lamps, whale oil was beyond the reach of merchant sailors and of nearly everyone else. In the 1850s and for decades thereafter, whale oils and especially spermaceti commanded high prices for their use as lubricants; the buyers of American whale oil were the owners of industrial production plants in England and Germany. The value of exported whale and spermaceti oil exceeded a million dollars a year through the 1850s and 1860s. Nearly all the exported whale oil went through New York harbor.

Whale oil had created a market for a safe, clean, and bright-burning lamp oil. As the price of the oils of whales rose, an alternative fuel appeared. Abraham Gesner, a physician and mining hobbyist from Newfoundland, isolated and purified a new kind of lamp oil. Gesner had developed and refined the lamp oil he named "kerosene" in Newfoundland, but he moved to New York to campaign for its production.

Gesner and family arrived in Williamsburg (now a neighborhood of Brooklyn) just as "Moby-Dick or the Whale" was finding its way into readers' hands. Gesner was welcomed by both scientific and investment communities of New York. The production company that Gesner helped form and whose factory Gesner designed and developed, would come to be known, after repeated restructurings and changes in ownership, as the "New-York Kerosene Oil Company." The company's presence in New York was like that of a bright meteor: sudden, surprising, brilliant, and brief.

P. Spellane, *Chemical and Petroleum Industries at Newtown Creek*,
History of Chemistry, https://doi.org/10.1007/978-3-031-09629-7_5

Abraham Gesner had been born to a farm family in Nova Scotia in 1797 and educated near his home [1, 2]. His recent ancestors had emigrated from the Netherlands and established a home in Tappan, New York, on the west side of the Hudson River, about 20 miles north of Manhattan. Abraham Gesner's father fought in the Revolutionary War, but, as a citizen of the United States may be inclined to think, on the wrong side. As the war concluded, the senior Gesner moved the family to the north, to build a home among British loyalists in Nova Scotia.

As a young man, not unlike his contemporary Martin Kalbfleisch, Gesner's first venture from home was as a merchant sailor, a career that lasted a few years, as did that of Kalbfleisch. Gesner returned to the family farm in Nova Scotia and married, at age 27, the daughter of Nova Scotian physician. With new wife, Gesner departed Nova Scotia for London where he studied medicine at St. Bartholomew's Hospital and surgery at Guys Hospital. He earned degrees in both medicine and surgery, but more importantly, he made the acquaintance of Charles Lyell. Lyell had published "Principles of Geology: being an attempt to explain the former changes of the Earth's surface, by reference to causes now in operation, in three volumes" in the early 1830s, several years before Gesner arrived at St. Bartholomew's. As his acquaintance with Lyell grew, Gesner began to understand the principles and practices of geology.

On returning to Nova Scotia, armed with two medical degrees, Gesner appears to have taken up the storybook life of a country doctor, but his interest in geology grew as both a source of delight and distraction. His career reveals Gesner to be curious, observant, and thoughtful, an autodidact with the mind and hands of an engineer. Even as he went about his doctoring, Gesner found great satisfaction in finding and documenting the island's hills and forests and especially its rocks.[1]

As he would soon become with coal, Gesner was intrigued by tar and tar sands, and he was not alone in being so.[2] Gesner took a particular interest in an oil-rich coal that he discovered in his field work at New Brunswick, perhaps with the hope of identifying a native mineral that might provide economic benefit to the area. Despite his failure in an earlier attempt to encourage mining at Nova Scotia, Gesner studied the New Brunswick coal. At his home laboratory, Gesner succeeded in isolating from the coal a liquid that burned with a bright flame. He was inspired.

Gesner's interest in the components and economic potential of legendary tars of Barbados and Trinidad began in his early days as a seaman; he wrote that he had visited the Barbados tar pit in 1818. There are legends of the young Gesner's seeking to trade North American horses in the Caribbean, including his being shipwrecked twice in attempts to do so. One can imagine that, when Gesner, having succeeded in extracting value from the Albert mineral, met Thomas Cochrane (1775–1860),

---

[1] Gesner published on the subject; his work *Remarks on the Geology and Mineralogy of Nova Scotia* was printed by Gossip and Coade, Times Office, Halifax, Nova Scotia in 1836.

[2] Gesner had developed a friendship with Thomas Lord Cochrane, the 10th Earl of Dundonald, on the visit of the latter to Nova Scotia. On one occasion Gesner and Dundonald pursued an examination of tar-sands in Barbados. Professor Benjamin Silliman, Sr., who established the study of chemistry at Yale University, published notes describing tars discovered in New York and Pennsylvania. Several decades later, his son, Benjamin Silliman, Jr., provided a chemical analysis of Pennsylvania "rock oil.".

Tenth Earl of Dundonald and Admiral of the British North American and West Indian Station,[3] at that time headquartered in Halifax, the two would strike up a friendship and soon speak of pulling riches out of tar. Described as model for Lord Byron's Don Juan and having recently recovered his good standing in the Royal Navy, Dundonald was serving the final years of a long, legendary Naval career when he met Gesner at Halifax.

Gesner gave the first few of many lectures on coal-tar gas and coal oil. Encouraged in his attempts to extract oil from the Albert mineral, and ready for a new project, Gesner had just begun to speak publicly of the oil he called kerosene. Like Gesner, Dundonald had recurring dreams of the fortunes that tar held. In Dundonald's case, his tar-obsession was genetic: Dundonald's father had patented a method for distilling coal tar for illuminating oil, waterproofing material, and other uses.[4] As might happen if an idealist discovers that one of his hero's shares his own ambitions, Gesner was moved to dream with Dundonald of adventures in exploiting what they both believed to be the boundless, wealth-generating components of the tar in Trinidad. From Beaton [1]:

> We know that during 1851-1853 Dundonald filed in England a series of comprehensive patents covering in the broadest terms the applications of natural asphalt. These patents were confined chiefly to the use of asphalt as a paving material, as a mastic, as a "hydraulic concrete" suitable for fashioning into water pipes and sewer mains, and as an insulating material for electric wires.

Gesner, the ceaseless student, took note of Dundonald's approach. Even as he and Dundonald resumed their separate journeys, each seemed recharged by the time spent with the other.

Sometime in the early 1850s, evidently encouraged by his success in recovering oil from the native mineral and perhaps with a sense of mission he shared with Lord Dundonald, Gesner took the extraordinary step of moving his family to New York. He found a home in Williamsburg in northern Kings County, across the East River from Manhattan and a kilometer south of Newtown Creek. In New York, Gesner sought support for the commercialization of the Nova Scotia coal. He envisioned a significant market for the bright-burning coal extract, and he believed New York City to be the place where bright ideas gain support.

Gesner readily established a presence in New York City. He offered lectures to science-minded audiences in the city. An October 8,1853, a report in Scientific American [3] describes a presentation by Gesner in which he introduced his several kerosene illuminants. The journal article reports, "The passing of air through naptha and benzole fluids, thereby impregnating it with carbon and hydrogen, in the proper quantities for producing a bright light is nothing new; but hydro-carbon fluids produced from the asphaltum, and applied for this purpose, is a novel application."

---

[3] Dundonald's life is well-documented; biographical works include Admiral Lord Cochrane, The Autobiography of a Seaman (London, 2 vols., 1860–61); The Life of Thomas, Lord Cochrane, 10th Earl of Dundonald, G.C.B., by Thomas, 11th Earl of Dundonald and H. R. Foxbourne (London, 2 vols., 1869); and an obituary in Scientific American, 24 Nov. 1860.

[4] English patent No. 1291 (1781).

The article mentions a company having been formed for developing Gesner's invention of fluids obtained from the New Brunswick asphaltum; the writer extended to Gesner all best wishes for success.

Among Gesner's early supporters in New York was Horatio Eagle, one of several investors who joined Gesner in developing a new business venture. Eagle issued a printed pamphlet that described the new New York corporation and invited investors [4]. Although Gesner had not yet secured a US patent, the corporation documents specify the products the company would secure from the asphaltic mineral of Albert County, New Brunswick. "The Asphalte Mining and Kerosene Gas Company" was incorporated in New York County in June 1853 [5].

The founders of the new corporation resisted limiting the possible businesses of the new venture. It seems they agreed on short list of possibilities; the corporation evidently attracted investors, and the officers quite promptly sought to establish a site for operations. Articles 1 and 2 of the corporation's founding document provide the idea:

> First The Corporate name of the Company hereby formed shall be "The Asphalte Mining and Kerosene Gas Company"

> Second The objects for which such Company is formed are 1st the mining of Asphaltum Asphalte rock and other minerals 2nd the manufacturing of illuminating gas, burning fluids &c, and 3rd the manufacture of mastics cements and hydraulic concretes.

Securing rights to a production site and construction of a minerals processing plant were the new company's next and more daunting tasks. Both were accomplished with speed. In April 1854, would be developers Neziah Bliss and Samuel Sneden sold seven acres of undeveloped farmland along the northern shore of Newtown Creek in three tracks to Horatio Eagle, Erastus Smith, and Philo Ruggles, all of whom reassigned their tracts to the North American Kerosene Gas Light Company in May [6].

Construction of the kerosene refinery began almost as quickly [1, 2, 7, 8]. Gesner's patents appeared a year after the investment group purchased property on Newtown Creek and Eagle circulated optimistic advertising brochure. Construction at Newtown Creek began. Beaton reports [1]:

> Gesner designed the buildings and equipment; Stillman and Allen, doing business as the Novelty Iron Works, foot of 12th Street and East River, fabricated the distilling apparatus to his specifications; the chemicals used to treat the product were supplied by Martin Kalbfleisch of Bushwick; and the raw material was a cannel coal very similar to the Albert County mineral, from Dorchester, Westmoreland County, New Brunswick, across the bay from Albert County.

In June 1854, the first of Gesner's US kerosene patents was published with rights assigned to the Asphalte Mining and Kerosene Gas Company (Horatio Eagle witnessed Gesner's patent) [9].

A year later, several "improvement in kerosene" patents were published [10–12]. In describing his process for isolating kerosene from asphaltic minerals, Gesner describes dry distillation as the first step, roasting the mineral somewhat gently to collect volatile gases, used for lighting or as fuel for the heating steps, and to collect liquid condensate. The initially formed liquid comprises water, ammoniacal liquid, and tar, all of which are more dense that the lighter oil that contains the product. The lighter phase liquid is isolated and distilled a second time, again at temperatures low enough to prevent decomposition of its kerosene content. Gesner describes treatment of the distillate [10]:

> The light liquid is transferred from the receiver to a suitable vessel or vat and mixed thoroughly with from five to ten per cent. of strong sulphuric, nitric, or muriatic acid, according to the quantity of tar present. ... I have enumerated three acids, but I give the preference to sulphuric, although either of the others will answer very well. I also mix with the liquid one to three per cent. of peroxide of manganese, according to the turbidness of the liquid, about two per cent. being the average quantity required.

And later in the treatment of distillate [10]:

> I next mix the distillate with about two per cent., by weight, of powdered and freshly calcined lime. The latter, by its powerful affinity for water, will absorb it thoroughly from the liquid hydrocarbon which always at this stage of the process contains it in greater or less quantity. Lime by its alkaline properties will also neutralize any acid in the liquid.

Through a quick succession of reorganizations, the original Asphalte Mining and Kerosene Gas Company became the North American Kerosene Gas Oil Company and, later, the New-York Kerosene Company, the country's largest-scale refiner of coal oil and proprietor of the name "kerosene" and owner of the large refinery at Newtown Creek.[5]

New-York Kerosene (Fig. 5.1) was the center of expertise in the art of extracting kerosene from coal, but coal could not compete with petroleum. The Kerosene Oil company would, within about a decade, be declared bankrupt. Abraham Gesner was by that time well-separated from the company. As the company resolved its few production issues, the self-titled Colonel Edwin Drake developed a method for extracting petroleum from its seemingly limitless subterranean pools. Early one morning in 1859 in western Pennsylvania, petroleum began to displace coal oil. In the early 1860s, in that moment of coal oil's slow decline and rock oil's rapid growth, Gesner published a book on both oils and the fractions distilled from them [8].

---

[5] The original Asphalte Mining and Kerosene Gas Company was reborn as the North American Kerosene Gas Company not long after the ink dried on the Asphalte Mining documents, reflecting the company's greater interest in kerosene and diminished interest in mining. Later still, the North American Kerosene company reorganized as the New-York Kerosene Oil Company. We have not found charter documents for either of two successors to Asphalte Mining company, but it appears that in each change, the principals attached to the company, and as suggested by Daum and Williamson [13], the chief engineer and production methods, and eventually sulfide-removal strategies underwent changes and improvements. In its final, longest-lived and most successful form, the company was called the New-York Kerosene Oil Company (Fig. 5.1). This string of companies (Asphalte Mining, North American Kerosene, and New-York Kerosene Oil) had rights to Gesner's patents, any of these and only these could be called "the Kerosene Company." Early in its history, the Kerosene company gained rights to the intellectual property assets of the Samuel Downer Company of Boston. Downer had acquired exclusive rights to James Young's claims, which were slightly older and arguably dominated Gesner's. Downer and the Kerosene company settled their differences rather quickly and got along well for most of the coal–oil period.

Fig. 5.1 The site of the New York Kerosene Oil Company on Newtown Creek, 1864 (Courtesy of The New York Public Library [14])

Loyalty to the notion that the only genuine kerosene is that which is extracted from coal and to the original Kerosene Oil company (and its direct descendants) proved to be strong. Reports, advertisements, and legal announcements concerning the New-York Kerosene Oil Company published in *The Brooklyn Daily Eagle* provide a running view of the company's rapid rise and slow decline:

April 22, 1857, Advertisement: "KEROSENE OIL—NOT EXPLOSIVE.—the advantages possess by the Kerosene Oil are: 1st The intensity of the light produced, 2nd It is not explosive. 3rd It will remain fluid when best sperm oil has congealed. 4th Its unrivaled economy--$4.10 worth Kerosene Oil gives as much light as $9 of rapeseed oil --$12 of whale--$26.47 sperm oil—or $29 of burning oil. This Oil can be seen burning at all times, day and evening, at our store, 62 Fulton Street, Brooklyn."

July 30, 1859, Reporting: "More oil is made from coal in one week than was ever obtained by our whale-fishers in the most fortunate season they ever enjoyed. The oldest factory of this character in American is the Kerosene Works on Newtown Creek near the Eastern District of Brooklyn, on Long Island, and owned by the New York Kerosene Oil Company, which has control of the two richest oil-bearing coals that have yet been discovered. ...

"Lately, however, there have been erected several rows of retorts, each of which contain 25 tons of coal, and this amount is worked off, as a regular charge, at one operation. ... Each of these retorts is built of brick, in the form of a huge pipe bowl, and, when the coal is packed in, the fire is kindled on the top with anthracite, the downward draft of heat is effected by steam power, and the oil vapors that are carried off below are condensed into crude-oil, and pass from a conducting pipe into a tank. ...

"There is now in course of erection at these works a retort such as we have just described, which will be able to smoke 100 tons of coal at a time."

September 9, 1859, Reporting: "Kerosene Oil": "Furnishing a light nearly equal to the best coal gas, ch(e)ap, clean, perfectly sa(f)e and free from any unpleasant odor, it is a boon of no little value. The Kerosene Oil was soon appreciated, and the demand for the article has been unprecedented; so rapidly did orders pour in from all parts of the country that the limited extent of the works first established did not enable its owners to complete half of them. But the New York Kerosene Oil Company made prompt efforts to be able to meet the demand, by the construction of works on an adequate scale, and their works on the Newtown Creek, near Greenpoint, is one of the most extensive and complete establishments. They have the sole right to manufacture this kind of coal oil in the United States, and the perfection to which they have carried this article bids fair to drive all other description of coal oil out of the market.

"The ground covered by the works has an extent of fourteen acres, and the works and machinery, all very peculiar in their nature the result of experiments that have cost nearly one hundred thousand dollars, are on the most extensive scale. They manufacture now about 6,000 gallons of oil per day, but this may be increased. The process of manufacture is thus described:

'There are eighteen large pipes of peculiar form. They are twelve feet in diameter and twelve feet deep, formed of a wrought iron shell covered on the outside with brick, and lined with fire brick. At the bottom is placed a grate, and from the bottom a chimney comes out, and extends upwards at the side of the 'pipe' to a short distance above the top. Near the top of this chimney a steam jet enters for the purpose of making a draft. Into these 'pipes' which are named 'meerschaums,' from their resemblance to a tobacco pipe, are pl(ac)ed twenty-five tons of bituminous coal. On the top of this twenty-five tons of coal is placed three tons of anthracite coal. All is then ready and the anthracite is set on fire, and the process goes on. The draft is made by means of the steam jet which forces a jet of steam into the chimney and thus causes a draft sufficient to keep the combustion going on. The air enters freely at the top of the 'meerschaum,' passes though the burning anthracite, by means of which it is deprived of the oxygen, and of all gases which would have a tendency to have an injurious action on the oily vapors. The coal is thus decomposed by the hot gasses passing through it, and the vapor is condensed at the bottom, when it passes off in the form of a dark brown colored oil. This operation consumes two days for this 'meerschaum.' The residue is nearly equal in bulk to the coal, and abounds in alumina containing but a very small amount of carbon. The amount of oil extracted is one hundred and ten gallons to the ton of coal. From the 'meerschaums' the oil passes off by means of pipes to a large reservoir, underground, capable of containing thirty thousand gallons, and from thence it is conveyed to another part of the works for distillation." The oil when ready for market is a clear colorless fluid, with scarcely a perceptible smell, and non-explosive."

October 22, 1861 and frequently repeated until January 22, 1862, Advertisement: "New York Kerosene Oil – The Price of our standard KEROSENE ILLUMINATING OIL has been still further reduced to meet the market. Although the oil is superior to any other it will be sold at the same price as the explosive oils now in general use. CHAPPEL & POOL, Wholesale Agents, 64 Fulton Street.

Annual reports indicate the growing debts of the Kerosene company as competition from the petroleum producers undermine the profitability of the Kerosene company:

January 22, 1862, Reprinted: THE NEW YORK KEROSENE OIL COMPANY, (a Corporation duly created and organized under the Act of the Legislature of the State of New York), Annual Report is quoted: the amount of the capital of said company is $500,000, and the full amount thereof has been paid in, which includes $25,000 in cash and $475,000 in stock issued for certain property, purchased by the Trustees and conveyed to the Company; debts $22,282.13.

January 20, 1865 Reprinted: THE NEW YORK KEROSENE OIL COMPANY, (a Corporation duly created and organized under the Act of the Legislature of the State of New York), Annual Report is quoted: the amount of the capital of said company is $500,000, and the full amount thereof has been paid in, which includes $25,000 in cash and $475,000 in stock issued for certain property, purchased by the Trustees and conveyed to the Company; debts $27,479.72.

September 10, 1868, Notice: "IN BANKRUPTCY – Eastern District of New York, ss. At the City of Brooklyn, the 8th day of September, A.D. 1868. The undersigned hereby give notice of his appointment as assignee of The New York Kerosene Oil Company. of Newtown, in the County of Queens and State of New York, within said district, who has been adjudged a bankrupt upon his own petition by the District Court of Said District. To CHARLES JONES Assignee, &c"

October 22, 1868, Notice: "U.S. District Court, Eastern District of New York. – In the matter of the New York Kerosene Oil Co., in Bankruptcy. ASSIGNEE'S SALE – Notice is hereby given, that I will sell at Public Auction, on FRIDAY, 30th day of October, 1868 at 11 o'clock A.M., at the Manufactory of said Company, situate on Newtown Creek, near Greenpoint ave. in the town of Newtown, L. I., about 1175 tons of coal, 2000 new oil barrels, oil, acids, paint, iron, truck, harness, canal boats, barges, bulk boats, pumps, machinery, &c., &c., property of said Bankrupt Corporation. Dated Brooklyn, Oct. 19, 1868. CHARLES JONES, Assignee in Bankruptcy."

February 8 and 9, 1869 and notices of reschedule to February 12 and then February 13, 1869: JAMES COLE'S SON, AUCTIONEER, Wednesday, Feb. 10, 1869, at 11 o'clock A. M., at the New York Kerosene Company's Works, on Newtown Creek, near Greenpoint ave. Assignee's Sale of Lot Scrap Iron, Pipe Fittings, Valves, Lathes, Iron Tanks, Oil White Lead, Coopers' Tools, &c. property of said bankrupt Corporation. By order of CHARLES JONES, Assignee.

December 21, 1870, Notice: BY JAMES COLE'S SON. J. COLE, Auctioneer. Office No. 389 (old No. 362) Commercial Exchange, opposite City Hall, Brooklyn. Thursday December 22, 1870, At 1 o'clock A.M., on the premises lately occupied by the New York Kerosene Oil Company, on Newtown Creek, near Greenpoint avenue. Part of the assets of said Company, consisting of large and small engines, Worthington and other pumps, oil tanks, stills, steam boilers, &c. &c. Terms cash and deposit required. Catalogues at the office of the auctioneer, No. 389 Fulton st. Brooklyn. Sale by order of CHARLES JONES, Assignee in Bankruptcy.

June 7, 1872, Notice: J. COLE, Auctioneer. Tuesday June 11, 1873. At 9 ½ o'clock A.M. at the New York Kerosene Oil Co.'s Works on Newtown Creek, in Queens County, a short distance from the Greenpoint avenue bridge. By order of the Assignee in Bankruptcy, a large quantity of old iron to be sold by weighmaster's certificate.)

October 24, 1873, Notice: J. Cole AUCTIONEER, Commercial Exchange, No, 389 Fulton Street, Opposite the City Hall, Monday, Oct. 27 at 11A.M. At the works of the late New York Kerosene Oil Company, now occupied by R. W. Burke, Esq., on the New Town Creek, near the Penny Bridge, One 350 horse power walking beam-condensing steam-boat engine, several large boilers. Worthington pumps, one 75 horse power engine, damaged; also about 300 tons of wrought, cast and scrap iron pipes, stills, etc. By order of Charles Jones, assignee of the New York Kerosene Oil Company. Terms cash.)

References

1. Beaton K (1955) Dr. Gesner's Kerosene: The Start of American Oil Refining. Bus Hist Rev 29:28–53
2. Gesner GW (1896) Dr. Abraham Gesner, A Biographical Sketch. Bulletin of the Natural History Society of New Brunswick, XIV:3–11
3. Anon (1853) New light - kerosene gas. Sci Am 9(4):29
4. Project for the Formation of a Company to Work the Combined Patent Rights (for the State of New York) of Dr. Abraham Gesner, of Halifax, N.S., and the Right Hon. The Earl of Dundonald, of Middlesex, England. The Bella C. Landauer Collection at the New-York Historical Society Library.
5. Document of incorporation, Old Records Office of the Clerk of New York County.
6. Deeds: Bliss and Sneden to Eagle, Queens County Land Records, Liber 128, p 263; Eagle to North American Kerosene Gas Light Co., p 346; Smith to Kerosene Co., p 344; Ruggles to Kerosene Co., p 341
7. Anon (1884) Kerosene – The origin of the name, the history of a great industry of years past, and possibility of its revival. Eng Min J 57(6):99–100
8. Gesner, A (1861) A Practical Treatise on Coal, Petroleum and Other Distilled Oils. Bailliere Brothers, New York, pp 8–9; drawings contained within are believed to be those of buildings and equipment at the Kerosene Works
9. Gesner, A (1854) Improvement in kerosene burning-fluids. U.S. Patent No. 11,205, June 27, 1854
10. Gesner A (1855) Improvement in processes for making kerosene. U.S. Patent No. 12,612, March 27, 1855
11. Gesner A (1855) Improvement in burning-fluids. U.S. Patent No. 12,936, May 22, 1855
12. Gesner A (1855) Improvement in burning-fluid compounds. U.S. Patent No. 12,987, May 29, 1855
13. Williamson H, Daum A (1959) The American petroleum industry: the age of illumination 1859–1899. Northwestern University Press, Evanston
14. Lionel Pincus and Princess Firyal Map Division, The New York Public Library (1864) Higginson's plan of the city of Brooklyn, L.I. Retrieved from https://digitalcollections.nypl.org/items/c69f7a9c-2199-8e38-e040-e00a18066e4a

Chapter 6
Benjamin Silliman, Jr. and The Pennsylvania Rock Oil Company

There were two Professors Benjamin Silliman at Yale College, Silliman Senior and Silliman Junior. Father and son worked, travelled, lectured, and wrote, seemingly without pause, to a broad audience, each quieted only by death. Their careers defined the dimensions of professional chemistry in the United States: Silliman Sr. on higher education in chemistry and Silliman Jr. on the application of chemistry to commerce. The Sillimans shared an appreciation for the history of chemistry and concern for chemistry's role in society. They liked mines and minerals and, through their analyses of diverse mining sites and mineral samples, advanced the practice and profitability of mining. Both father and son were drawn to petroleum, the tar-like mineral resource. In 1855, Silliman Jr. published a concise, thorough analysis of a sample of petroleum collected at Oil Creek in western Pennsylvania. Silliman's report secured petroleum's place in the commerce of the United States [1].

Benjamin Silliman Sr. was Yale College's first professor of chemistry. When offered the position by Yale president Timothy Dwight,[1] Mr. Silliman was pleased indeed with the offer, but pointed out to President Dwight that he would need to gain some understanding of chemistry before he could accept the new position. Silliman Sr. had studied theology and rhetoric at Yale, studied law at law offices in New Haven, and had been admitted to the practice of law, but Silliman Sr. had no experience in chemistry. That Silliman would require time and opportunity to learn chemistry was agreed: his education began in Philadelphia and continued at each of the country's several centers of expertise followed by study at European universities. His education in chemistry consumed several years, and fitting a laboratory space with the tools, glassware, and the reagent chemicals with which professors of chemistry teach required more of President Dwight's time and money. Finally, having established his own approach to the study of chemistry and constructed a suitable venue for instruction, Professor Silliman, Sr. was ready.

---

[1] There were two Yale presidents Timothy Dwight, one grandson of the other. The elder Dwight offered the professorship to the elder Silliman.

P. Spellane, *Chemical and Petroleum Industries at Newtown Creek*, History of Chemistry, https://doi.org/10.1007/978-3-031-09629-7_6

Silliman Sr. launched the study of chemistry at Yale, but his influence on the practice of scientific observation and reporting was not confined to chemistry or to events at Yale. The scion of an old Yankee family, son of a Revolutionary War hero, the man educated for a life in theology and law, found his place as an observer, examiner, and instructor in physical and medical sciences. His appetite for observation and analysis was expansive, and his instinct to share what he learned was as strong as his curiosity. He wrote about science, travel, society, and art; he founded and edited the extraordinary *American Journal of Science and Arts*. The journal gave Silliman Sr. license to read, review, publish and comment on the work of many scientists. The American Journal came to be known as *Silliman's Journal* and its editor recognized as one of the young nation's leading scientific thinkers.

With the arrival of Benjamin Silliman Jr. in 1816, Silliman Sr.'s impact grew larger still. The Silliman apple fell not far from the Silliman tree. Benjamin Jr. grew up at Yale, earned a degree there, and made a career in New Haven, assisting his father in all his scientific work even as he pursued his own chemistry interests.

The younger Silliman built a new laboratory in disused space in the old Yale President's residence (for which he paid rent to Yale). A consummate consultant in several fields of expertise, Silliman Jr. was sought by both clients and students. As a consulting scientist, he conducted analyses for clients: the quality of water in a town's reservoir; the mineral content of specific mines in the western United States; the geology and arability of ranchland in southern California. As a private educator, he received students and taught chemistry. Yale eventually recognized Benjamin Jr.'s skill for attracting students and his success in extending practice of professional chemistry: in 1858, Yale appointed Silliman Jr. "Professor of Applied Chemistry" and appointed Silliman's colleague John Pitkin Norton "Professor of Agricultural Chemistry" [2].

An individual who maintains a career in science is likely to hold a sustained interest in a few reasonably well-defined areas of inquiry. In the breadth of their scientific interests, both Sillimans were outliers. Boundless curiosity is especially characteristic of Silliman Sr., whose *American Journal of Science and Arts* reads like a randomly grouped collection of observations, sightings, analyses, and thoughts about every physical observable phenomenon. A reader gets the idea that, in the mind of this journal's editor, every scholarly effort, examined and reported in earnest, is of importance to the advancement of the nation's "science and arts." Reports in the journal are written and arranged with the enthusiasm of a kid reporting on new and wonderful candy stores. An unending train of two- and three-page reports of observations and measurements of the natural world seem linked one to another by the journal editor's joy.

Silliman Jr. focused his effort on chemical phenomena (and medicine) and the application of chemistry knowledge and know-how to matters that affect the general economy. He was consulted on subjects such as water quality and the accessibility and potential value of sites and their natural resources. Despite their several differences in emphasis and approach, the two Sillimans shared an interest in the availability, amount, physical properties, and potential uses of petroleum. Both express an appreciation for the traditional uses of tar, as for example by the Seneca natives.

Twenty-five years before there was a frenzy for petroleum, a frenzy that Silliman, Jr helped stoke, Silliman Sr. wrote a charming account of his journey with a friend to a site near the town of Cuba in the southern tier of western New York to observe the town's "Oil Spring." His "Notice of a Fountain of Petroleum, called the Oil Spring" appeared in the American Journal in 1833 [3]. Silliman tells readers the exact location of the site, describes the color, feel, and odor of the petroleum, speculates on its chemistry, and describes a few rudimentary processing steps.

Silliman Sr. begins by describing what he understands of practice of people living near the spring [3]:

> They collect the petroleum by skimming it, like cream from a milk pan; for this purpose, they use a broad flat board, made thin at one edge, like a knife; it is moved flat upon, and just under the surface of the water, and is soon covered by a coating of the petroleum, which is so thick and adhesive that it does not fall off, but is removed by scraping the instrument on the lip of a cup. It has then a very foul appearance, like very dirty tar or molasses but it is purified by heating it and straining it, while hot, through flannel or other woolen stuff. It is used, by the people of the vicinity, for sprains and rheumatism, and for sores on their horses, it being, in both cases, rubbed upon the part.

He moves on to speculate the chemistry of the petroleum and its companion gas [3]:

> We had no means of collecting [gas that is constantly escaping through the water, and appears in bubbles upon its surface] or of firing it, but there can be no doubt that it is the carburetted hydrogen, – probably the lighter kind but rendered heavier and more odorous by holding a portion of the petroleum in solution... We could not learn that any one had attempted to fire the gas, as it rises, or to kindle the film of petroleum upon the water: it might form a striking night experiment....

> As to the geological origin of the spring, it can scarcely admit of a doubt, that it rises from the beds of bituminous coal, below; at what depth we know not, but probably far down; the formation is doubtless connected with the bituminous coal of the neighboring counties of Pennsylvania and of the West, rather than with the anthracite beds of the central parts of Pennsylvania.

> A branch of the Oil Creek, which flows into the Allegany River, a principal tributary of the Ohio, passes near this spring....

Silliman Sr. speculates whether petroleum samples recovered at the spring near Cuba, New York share a geological history with that of the oil collected along "Oil Creek, which empties into the Allegany River in the township and county of Venango" [3]:

> The petroleum, sold in the eastern states, under the name of Seneca oil, is of a dark brown color, between that of tar and molasses, and its degree of consistence is not dissimilar, according to the temperature; its odor is strong and too well known to need description.

> I have frequently distilled it in a glass retort, and the naptha which collects in the receiver is of a light straw color, and much lighter, more odorous and inflammable than the petroleum; in the first distillation, a little water usually rests in the receiver, at the bottom of the naptha; from this, it is easily decanted, and a second distillation prepares it perfectly for preserving potassium and sodium, the object which has led me to distil it, and these metals I have kept under it (as others have done) for years; eventually they acquire some oxygen, from or

through the naptha, and the exterior returns, slowly, to the condition of alkali – more rapidly, if the stopper is not tight.

Noting that "the petroleum remaining from the distillation, is thick like pitch," Silliman considers the possible common origin of petroleum collected along Oil Creek or at the Oil Spring presently considered and the "maltha and petroleum, in the island of Trinidad" and the connection of either or both to bituminous coal, and the related question: does petroleum on the surface of a creek or spring imply the certain presence of coal beneath the surface? Silliman Sr. makes an eerily prescient suggestion to landowners who suspect coal lies beneath the surface petroleum: "It would be much wiser *to bore*; which would enable them, at a comparatively moderate expense, to ascertain the existence, depth and thickness of the coal, should it exist."

Twenty years later, Silliman Jr. would have occasion to examine petroleum collected in Venango County, Pennsylvania for its inherent value, not solely as an indicator of bituminous coal. Why the several investors and directors of the Pennsylvania Rock Oil Company sought Silliman Jr.'s analysis of Pennsylvania "rock oil" is clear. Silliman Jr. was Professor of Applied Chemistry at Yale, had built a career as a scientific consultant, had examined and reported on the chemistry of drinking water from particular sites, and published analyses of the mineral wealth of specific mines in California. He had a record for successfully studying situations, answering questions, and producing reports on time.

George H. Bissell [4] had studied at Dartmouth and graduated in 1845. He taught in secondary schools in New England, sought more rewarding opportunities in Washington DC and in New Orleans where he became Superintendent of Schools. He continued to study even as he pursed teaching responsibilities in New Orleans, studying modern languages and law. With a law degree, Bissell moved to New York to begin a law practice. In the course a visit to his mother in Hanover, New Hampshire in 1854, Bissell called upon an acquaintance, Professor Dixi Crosby, in his office at Dartmouth. In the course of that visit, Bissell inquired about a petroleum sample that Dr. Francis Brewer had provided to Crosby.

Crosby told Bissell of Brewer's interest in the medicinal value of the creek oil collected near Titusville, Pennsylvania. Bissell seemed aware of new findings of illuminating oils obtained from coal[2]; he may have read the enthusiastic report in *Scientific American* [6] of a demonstration at the Art Union in New York of light produced in the combustion of two variations of Gesner's "kerosene" or known of the coal oil company chartered at office of the New York County Clerk. Bissell studied the rock oil in the glass jar in Professor Crosby's office at Dartmouth and thought about petroleum and light [7].

On returning to New York, Bissell persuaded his law partner J. G. Eveleth to visit Venango County to determine if the area's rock oil, the dark liquid he had examined in Professor Crosby's office, might be adequate in supply to support its commercial development. Eveleth's findings were affirmative and optimistic; Bissell and Eveleth purchased or leased about 200 acres of land a few miles south of Titusville Pennsylvania. The two men organized the Pennsylvania Rock Oil Company of New

---

[2] In his book *Petrolia*, Brian Black reports that Bissell worked with in the coal oil business [5].

York. Papers for the joint stock company were filed in New York on December 30, 1854 [8].

It had been only weeks since Bissell first held a sample of the Venango County rock oil, but ambition and petroleum inspired quick decisions. Raising capital for the new venture took more time. Investors were not as quickly convinced as Bissell and Eveleth had been. A group of interested persons in New Haven, led by James Townsend, the president of a local savings bank to whom Bissell had been introduced, suggested that Bissell provide a sample of the rock oil to his New Haven neighbor Benjamin Silliman. The younger Silliman worked as a consultant and held research space on the Yale campus, but he was not yet a member of the faculty. It is reasonable to speculate that the New Haven business community would have been aware of the two Sillimans' interest in mining, natural oils, analytical chemistry, and, perhaps most importantly, commerce.

Bissell provided a sample of the Venango County oil to Professor Silliman for chemical analysis and recommendations on the possible uses to which the oil could be applied, including particularly as a source of illuminating oil. Silliman's analysis was clear, detailed, and unequivocal. When the Pennsylvania Rock Oil Company directors managed to raise the $526.08 Silliman charged for his work, Professor Silliman released the report.[3] Silliman's analysis, a model ten-page lab report, set in motion decisions and events that altered the direction of the country's economy, then the entire developed world's economy, and eventually the carbon cycle of the planet. A more immediate result was a new listing in Trow's Directory of New York City (1856): "Eveleth, Bissell, and Reed, oil, 346 Broadway, NY."

It is unlikely that one 10-page report [10] could change the course of human history, but the energy density of fossil fuels, the seductive warmth and brightness of refined petroleum's flame, and other qualities that Silliman suggests in his short report have made rock oil difficult to resist.

The Pennsylvania Rock Oil Company was reorganized in late 1855 as a Connecticut company to address concerns of the New Haven investors. Bissell's company was by that time armed with Professor Silliman's analysis of the Venango rock oil. The company had attracted new investors; Benjamin Silliman was one.

Professor Silliman began his analysis with a 304 g sample; he used standard distillation technique to isolate 8 fractions, totaling 160 g and noticed that densities of the components grew slowly from 0.733 g/milliliter for the lowest boiling fraction to 0.854 for the highest boiling fraction. Silliman offers one other most significant insight: he suggests that the distillation itself may have altered the composition of the crude oil to produce the products that he had identified.

Seeking more valuable distillates, Silliman employed a copper still to enable distillation at very high temperatures. About 560 oz (14 imperial quarts) of the crude oil were heated to approximately 280 °C. The high temperature distillation produced 130 oz of "a very light colored thin oil, having a density of 0.792;" at 300 °C, 123 oz of a more viscous yellow oil of density 0.865. Silliman collected more samples at

---

[3] Daniel Yergin begins his analysis of the petroleum industry [9] with the anecdote of the unpaid bill.

the boiling temperature of mercury (357 °C), 170 oz of a dark brown with a strong "empyreumatic odor" which color and odor could be removed by wash with water to produce a bright burning oil.

Silliman makes clear that his exhaustive distillation of the crude oil may have wrought change in the material. He suggests that the waxy solid "paraffine" and some of the liquids collected at very high distillation temperatures were not native to the crude oil. He recognizes an effect that would become an essential part of petroleum refining, the processes related to "petroleum cracking," in which high molecular weight hydrocarbons in crude oil are purposely fragmented to form lower molecular weight alkanes and alkenes.

In describing the chemical properties of the distillates, Silliman reported that the oils are thermally stable and air-stable; they consist of hydrocarbons, compounds comprising only carbon and hydrogen in a ratio of slightly more than two hydrogen atoms for each carbon atom. The small aqueous content of the crude oil was easily removed; the oil appears to have no corrosive effect on clean copper and does not tarnish clean potassium, suggesting its oxygen-free composition. Both effects indicate the oils' suitability as lubricants for machinery.

Most compelling is Silliman's final assessment [10]:

> The result of this graduated distillation, at a high temperature, is that we have obtained over 90 per cent. of the whole crude product in a series of oils, having valuable properties, although not all equally fitted for illumination and lubrication…. It is safe to add that, by the original distillation, about 50 per cent. of the Crude oil is obtained in a state fit for use as an illuminator without further preparation than simple clarification by boiling a short time with fair water.

Silliman's inference from the results of fractional distillation is profound. His assessment of the sample he analyzed is both consistent with present day understanding of the composition of petroleum and prescient in proposing a reasonably straightforward method of its refining. In his analysis and assessment of the Venango County petroleum, Professor Silliman uses simple language to describe the chemistry and chemical technology upon which is built the industry that has determined an enormous part of global economics and politics for 150 years [10]:

> We infer from (more precise determinations of the boiling temperatures of each previously obtained fraction) that the Rock Oil is a mixture of numerous compounds, all having essentially the same chemical constitution, but differing in density and boiling points, and capable of separation from each other, by a well-regulated heat.

Edwin Drake was introduced to the Pennsylvania Rock Oil Company by way of the New Haven community.[4] The idea that one might drill for subterranean petroleum as one would drill for water was not Drake's alone; stories of recovering medicinal rock oil from water wells were legend. What exactly were the Rock Oil investors' instruction to Drake, who guided Drake in the art of constructing water-wells, how

---

[4] Brian Black [5] provides detail on the relationships, the persons involved, the locations, and the various ambitions of the Oil Creek community before and after word spread of Drake's successful oil rig.

he selected his site: reports on all these vary. What is clear is this: in 1859 Drake succeeded in constructing a petroleum drilling rig; he chose a suitable site and hit oil.

References

1. Papers of the two Professors Silliman are maintained at the Yale University Archive
2. John Pitkin Norton published "Elements of Scientific Agriculture". Biographical information on Norton is found in Haynes W (1954) American chemical industry, a history in six volumes, vol 1, a new industry: fertilizers, p 338 and Dict Am Biog XIII, 574 and Browne, J Am Chem Soc, 48, 177 (1926)
3. Silliman B (1833) Notice of a fountain of petroleum, called the Oil Spring. American Journal of Science and Arts 23:97–102
4. This description of Bissell, the Pennsylvania Rock Oil Company, and Seneca Oil Company follows Arnold R. Daum and Harold F. Williamson's encyclopedic history of the petroleum industry: Daum A, Williamson H (1959) The American petroleum industry, 1859–1899, the age of illumination. Northwestern University Press, Evanston, p 64
5. Black B (2000) Petrolia, the landscape of America's first oil boom. The Johns Hopkins University Press, Baltimore, p 28
6. Anon (1853) New light - kerosene gas. Sci Am 9(4):29
7. See interview with George H. Bissell, in Hayes SS, (1866) Report of the United States Revenue Commission on Petroleum as a Source of National Revenue, Special Report No. 7, 39th Congress, 1st Session. House Ex. Doc. 51:4–5
8. The charter of the Pennsylvania Rock Oil Company is filed at the Old Records Office of the Clerk of New York County
9. Yergin D (1990) The prize, the epic quest for oil, money, and power. Simon & Schuster, New York
10. Silliman B (1855) Report on the rock oil, or petroleum, from Venango Co. Pennsylvania. J H Benham's Steam Power Press, New Haven

Chapter 7
Charles Pratt, Henry Rogers, and Astral Oil

And this is luck for *me*. You and I are a team: you are the most useful man I know, and I am the most ornamental. Good-bye—I am feeling gay. I had nearly forgotten how it felt. Yours ever sincerely SLC

The conclusion of a letter from Samuel Clemens to Henry Rogers, December 21,1897[1]

Charles Pratt was born on a farm in Wilbraham, Massachusetts in 1830; he was one of eleven children. The beneficiary of education at Wilbraham Academy, Pratt was, from an early age, inclined for a life in commerce and drawn to New York City. An ambitious 21-year-old ready to build a career in the business of buying and selling, Pratt found employment in New York City and a home in Brooklyn.

Pratt's career in New York City would inspire any ambitious transplant seeking life and fortune in Gotham. Frederick Pratt provided a concise history of his father's preparation for commerce and public life [2]:

At the age of thirteen, he went to Boston as a clerk in a grocery store. A year at Newton, as an apprentice in the machinist's trade was followed by a year of formal education at Wilbraham Academy where he managed to live on a dollar a week. His business career really began when he engaged as clerk in a paint and oil shop in Boston, where he remained three years.

In New York, Pratt's first job was in the office of James and Daniel Schanck and Augustus Downing, "agents for French zinc paints and importers of French glass, paints, oils etc.,108 Fulton".[2] Within a few years, Pratt identified himself in city directories: "Pratt Charles, clerk, 108 Fulton, h, B'klyn." A few years later, the space at 108 Fulton Street was Pratt's alone. Pratt gained the acquaintance of workplace neighbors Charles Raynolds and Frederick Devoe, who, like Pratt, were in the

---

[1] Samuel Clemens (Mark Twain) and Henry Rogers met one another in 1893 and developed a deep and long-lasting friendship. Their frequent correspondence has been compiled, published, and is a pleasure to read [1].

[2] Information on Pratt's home and work addresses is gathered from information in New York City directories, published annually by several printers and available in the Library of the New-York Historical Society. Charles Pratt first appears in City directories dated 1851.

P. Spellane, *Chemical and Petroleum Industries at Newtown Creek*, History of Chemistry, https://doi.org/10.1007/978-3-031-09629-7_7

business of making and selling paint.[3] Pratt joined Raynolds and Devoe's company, reorganized as Raynolds, Devoe & Co. (1855), with operations at Pratt's office at 108 Fulton. Within two years, the company reorganized itself again as a partnership of three: Raynolds, Devoe and Pratt.[4]

The partnership was barely five-years old when Pratt proposed to his partners that they, paint-makers, could establish a commercial position in petroleum. It appears Pratt was impressed by the vigor of the new business of petroleum, and his partners impressed by Pratt's determination: the partners reorganized their professional ties. In 1864, the partnership that had been "Raynolds, Devoe and Pratt" became three separate companies: two were paint companies: C.T. Raynolds & Co. and F.W. Devoe & Co., and the third, an oil company, Charles Pratt Oil Works.[5]

How could Charles Pratt declare himself an "oil works?" We find no indication of Pratt's interest in petroleum that predates his establishing the Charles Pratt Oil Works, no evidence of his knowledge or involvement with oil, other than the oil in oil-based paints, but that provides a link. The patent record suggests that Pratt planned to reconfigure paint cans as kerosene-oil cans. It appears that, in common practice, paint-makers were, at the same time, makers of paint cans. To alert his partners to a position they could establish in the burgeoning kerosene trade, Pratt could point to their experience and know-how in can-design and manufacture. The paint-making partners could, working separately or together, make and sell kerosene cans to petroleum refiners or kerosene dealers.[6]

When Pratt had opportunity to build the "oil works," he made clear his grand ambition: he envisioned integrating petroleum refining and packaging. He could purchase kerosene from refineries and package the refined oil in cans labelled Charles Pratt Oil Works, or he could operate a refinery. As a refiner, Pratt would purchase

---

[3] Raynolds and Devoe were successors to an old and successful paint-and-pigment-making company started by the family of Gerardus Post in 1798. In the early 1820s, Post helped found one of the city's first chemicals manufacturers, the New York Chemical Manufacturing Company, which provided Martin Kalbfleisch employment in New York.

[4] The names and chronology of the various companies of Raynolds, Devoe, and Pratt are challenging. In describing a landmarked site in Manhattan, the New York City Landmarks Preservation Commission provides the following: October 28, 2008, Designation List 406 LP-2308, F.W. DEVOE & CO. FACTORY, 110–112 Horatio Street, Manhattan. Built 1882–83; Architects Kimball & Wisedell. Landmark Site: Borough of Manhattan, Tax Map Block 642, Lot 12. The Commission's footnote 21: "The full chronology of the firm's various names is as follows: William Post (1754–1798); William Post & Sons (1798–1800); William & Gerardus Post (1800–1834); William Post (1834–1835); Butler & Barker (1836–1846); Francis Butler (1846–1848); Butler & Raynolds (1848–1851); C.T. Raynolds (1851); Raynolds & Devoe (1852–1855); Raynolds, Devoe & Co. (1855–1858): Raynolds, Devoe & Pratt (1858–1864); C.T. Raynolds & Co., F.W. Devoe & Co. (two separate companies, 1864–1892); F.W. Devoe and C.T. Raynolds Company (1892-present day)".

[5] In 1892, Raynolds and Devoe reunited as The F. W. Devoe and C. T. Raynolds Company. After 1864, Devoe and Pratt directed separate business entities, but that did not prevent their joining forces in a subsequent venture.

[6] A report in the Cleveland Leader (September 1868) mentioned the Rockefeller, Andrews & Flagler company's (a predecessor of the Standard Oil Company, which was formed in 1870) contracting C. P. Born for 300,000 five-gallon cans to be used in shipping refined oil to Europe.

crude oil from "producers" (those who extract crude petroleum from the ground) then refine and package the various petroleum products in his proprietary cans.

To get his "oil works" venture underway, Pratt would seek out a reliable supply of kerosene, which led Pratt to Charles Ellis and Henry Rogers. Ellis and Rogers, natives of Fairhaven, Massachusetts, had joined the rush of the young and ambitious to the Oil Regions of western Pennsylvania. News in 1859 of Edwin Drake's oil rig led many to believe that wealth flowed from the ground along Oil Creek, an optimism that inspired Ellis and Rogers and many others to hie themselves to the oil fields. Just as reports of gold in the creeks of California had inspired a gold rush in an earlier decade, Drake's oil strike inspired an oil rush to western Pennsylvania.

Ellis and Drake, who as boys had likely spent more time on boats and docks laden with whale oil than on solid ground,[7] built and operated the Wamsutta Refinery at McClintocksville in Venango County. The two young men were of the generation of self-taught "refiners" of petroleum; their refinery in Pennsylvania's Oil Regions was close to what was thought to be the only source of petroleum production. Ellis and Rogers would purchase crude oil on location, refine it at McClintocksville, and market their refined product to vendors of kerosene.[8] Pratt, Ellis, and Rogers, three Massachusetts natives working far from home, made a deal: Pratt would buy, at an agreed price, the entire output of Ellis and Rogers's Pennsylvania refinery.

Their supplier-customer relationship fell apart when the price of Pennsylvania crude oil soared. Even in petroleum's early years, speculation was inherent and inevitable to the business. Ellis and Rogers fell victim to speculators' price manipulations; the two refiners discovered they could not purchase crude oil at the price they had agreed to sell refined oil. Their deal with Pratt was in tatters. Ellis abandoned the oil business, but Rogers did not. Rogers departed the Oil Regions for Newtown Creek, arrived at Pratt's place of business, and offered to fulfill his obligation to Pratt by working at the newly organized Oil Works. Rogers, one of the few experienced refiners of petroleum, and Pratt, the would-be marketer of well-canned kerosene, would join skills to build a new refinery in New York. Pratt, it appears, recognized and valued Rogers's refining experience and know-how. The two became partners in 1867, and their progress was swift: by 1868, the Pratt Oil Works had a petroleum refinery.

Pratt had established the Charles Pratt Oil Works in 1864, but the "oil works" did not involve refining until Henry Rogers arrived in New York. An 1868 insurance map provides details of the C. Pratt Oil Works, the locations of stills and barrel-making, at Blissville, on the Queens shore of Newtown Creek (Fig. 7.1). The 1868 map reveals more than it states: the label that reads "C. Pratt Oil Works Refused Permission to Make a Survey" appears to be pasted onto an earlier map: Pratt had acquired an abandoned oil refinery. He built the Pratt refinery on the site of New York's first and one of the country's largest kerosene refineries.

---

[7] Fairhaven, Massachusetts is a port along the Acushnet River, directly across river from the whaling industry's main port, New Bedford.

[8] Information provided in this section is found in histories developed by Christopher J. Richard, Director of Tourism for the Town of Fairhaven, Massachusetts, and the Fairhaven town library.

Fig. 7.1 Newtown Creek, East of the Greenpoint Avenue Bridge, including detail of the Charles Pratt Oil Works. From Higginson's Insurance Maps of the city of Brooklyn L.I. Surveyed, Drawn & Published by J. H. Higginson. 1868. Volume 4, detail from plate 85. Note that the Pratt Oil Works site is on the Queens (north, left in this presentation) side of the Newtown Creek (Courtesy of The New York Public Library [3])

Pratt had acquired rights to the site of Abraham Gesner's coal-sourced kerosene refinery. The New York Kerosene Works, the successor to original coal oil refinery constructed by Abraham Gesner in the early 1850s, had remained loyal to coal even as the petroleum industry ballooned in size and market share; by 1868, the site of the Kerosene Works had become available.

Newspaper accounts from the late 1860s chronicle the precipitous decline of the New York Kerosene Oil company. Having once been the country's largest refiner of coal oil and a key player in the country's kerosene economy, New York Kerosene Oil declined as quickly as petroleum advanced. The older company that had continued to extract kerosene solely from coal, even as petroleum proved itself a better source, was declared bankrupt in 1868.

Notices published in the *Brooklyn Daily Eagle* indicate the company's demise:

20 Jan 1865: THE NEW YORK KEROSENE OIL COMPANY, (a Corporation duly created and organized under the Act of the Legislature of the State of New York), Annual Report is quoted: the amount of the capital of said company is $500,000, and the full amount thereof has been paid in, which includes $25,000 in cash and $475,000 in stock issued for certain property, purchased by the Trustees and conveyed to the Company; debts $27,479.72.

10 Sept 1868: IN BANKRUPTCY – Eastern District of New York, ss. At the City of Brooklyn, the 8th day of September, A.D. 1868. The undersigned hereby give notice of his appointment as assignee of The New York Kerosene Oil Company. of Newtown, in the County of Queens and State of New York, within said district, who has been adjudged a bankrupt upon his own petition by the District Court of Said District. To CHARLES JONES Assignee, &c.

In bankruptcy the New York Kerosene Oil Company's property was assigned to Charles Jones; the map maker's relabeling the site "Pratt Oil Works" suggests that Pratt had purchased or leased the site.

With a new refinery under construction,[9] Pratt partnered with Devoe to market kerosene. Pratt and partner Henry Rogers refined petroleum; Pratt and former partner Frederick Devoe, would construct kerosene cans, then fill and market the product. As the two had done before, Devoe and Pratt set up a new company. The Devoe and Pratt Manufacturing Company was established in New York in 1868 with specific focus [4]:

> The objects for which the company is formed are for the manufacturing of cans and other vessels and for filling and packing oils and other fluids therein, and such operation as may be legally connected therewith under certain processes embraced in Letters Patent issued by the United States therefor to George W. Devoe, Frederick W. Devoe and Charles Pratt.

In constructing this new partnership, Devoe and Pratt would leverage their experience and growing patent position to make and fill good cans with exceptional kerosene. As venue for their new work, they chose Newtown Creek [4]:

> The operations of the Company are to be carried on in the Town Of Newtown, County of Queens, and in the City of Brooklyn, County of Kings and in the City and Country of New York, and the principal part of the business of said Company within this State shall be carried on in the Town of Newtown and said County of Queens.

About this time, Pratt secured proprietary position in this new business practice. Pratt was awarded two US patents with co-inventor Conrad Seimel (US Patents 87,704 (March 9, 1869) and 88,410 (March 30, 1869)) and three more as sole inventor (US Patents 89,167 (April 20, 1869), 96,958 (November 16, 1869), 98,408 (December 28, 1869)). All concerned the manufacture of oil cans, their shapes, their solders, and their nozzles.

The Devoe-Pratt joint venture did not last long: within a few years, as they had done before, the two dissolved this latest joint venture. Like an amicably divorced couple living in separate houses on the same street, Pratt and Devoe carried out operations in factory buildings on opposite sides of Newtown Creek, each removed a friendly but shoutable distance from the other.

In 1871 Henry Rogers secured a US patent that addressed kerosene's composition and safety; Rogers's assigning the patent to the Pratt company secured further advantage to their company. Rogers claimed improved methods for isolating and separating from kerosene the more volatile naphtha compounds in petroleum [5].

The flash point of a combustible liquid is the temperature at which the gas-phase concentrations of flammable material and oxygen are sufficient to support an explosive reaction. Naphtha's flash point is significantly lower than that of kerosene. Separating the more volatile naphtha from kerosene makes kerosene less prone to explosion; hydrocarbons with higher flashpoints are safer to handle. Eventually a market for refined naphtha would develop, but at the time of Rogers's invention, kerosene was

---

[9] It is reasonable to assume that the Kerosene Oil Company's refinery, designed for refining coal oil, could be reengineered for refining crude petroleum.

the valuable petroleum product. A technology that produced naphtha-free kerosene stood behind the reassuring claim that graced every can of Pratt's Astral Oil: "will not explode."

From Rogers's patent [5]:

> It is well known that when two or more liquids of different boiling-points are dissolved the one in the other, when heated so as to boil, a mixed vapor will be formed, partly of the lighter and partly of the heavier vapors. When this mixed vapor is condensed slowly, the heavier vapors are first condensed, mixed, however with a portion of the lighter. By repeating the operation of vaporizing and condensing many times an almost perfect separation of all the various products contained in petroleum many be effected. My invention consists in an apparatus for separating volatile hydrocarbons by repeated vaporization and condensation.

Rogers's patent details a formidable construction of hardware that enables isolation of volatile components of crude petroleum. The still comprises: a retort in which the application of high heat causes vaporization of the many components of petroleum, a multi-level "fractionating" column, and valves that direct mixed vapors to one of two chilled "worms" for condensation. In Rogers's design, the main condensing worm can be tapped at various points of its long travel, enabling an operator to collect distillates boiling in various narrow temperature ranges. Rogers's distillation apparatus consolidates the tasks of vaporizing components of a crude petroleum mixture and resolving "the mingled hydrocarbon vapors" into separate fractions, each characterized by a narrow range of boiling temperatures. The still separates the lighter hydrocarbons from the heavier and, working with the lighter "naphtha" hydrocarbons, resolves them into separate product types.

Scientific American published a report of the Pratt refinery in May 1872 [6]. At a critical moment in both the improvement of refining technology and the expansion of the petroleum industry, the Pratt refinery (Fig. 7.2) and its premium product Astral Oil stood as objects of technical expertise and performance. In their reporting on the early years of the American petroleum industry, Williamson and Daum refer to "Charles Pratt's greatly admired refinery in New York City" [7]:

> Pratt's refinery had a rated crude capacity of 1,500 barrels daily, which could yield a daily output of illuminating oils in excess of 1,100 barrels, or about 1,200 barrels of refined oils of all kinds, including naphthas. His main crude charging capacity was concentrated in a bank of four horizontal cylindrical stills, direct-fired, each charging 830 barrels. This total combined charge of 3,320 barrels could be made and run in the large stills twice weekly. The brick masonry surrounding the stills stopped at the center line, leaving the upper part exposed to the atmosphere in order to promote condensation of the heavy vapors and refluxing during cracking. Six small, horizontal cylindrical stills also took charges of 50 barrels each. In addition, Pratt had some small steam stills which he used exclusively to refine gasoline and naphthas.

Henry Rogers, Charles Pratt, and John D. Rockefeller had occasion to consider their mutual interests and competitive positions as they responded to formation of the South Improvement Company. The Rogers-Pratt-Rockefeller story concerns the confrontation between, on one side, petroleum refiners in New York and in the

Fig. 7.2 Pratt's Astral Oil Works (Image from 1872 Scientific American Supplement [6])

Oil Regions of western Pennsylvania, and, the other side, the South Improvement Company.[10]

In the period following the Civil War, the Pennsylvania legislature found cause to charter new commercial ventures as "improvement" companies which would, in principle, contribute to the post-war reconstruction. In chartering the improvement companies, the Pennsylvania legislature allowed some flexibility of intent. Improvement companies could comprise other companies, even companies chartered in other states, to participate in their shared goals of Improvement.

In 1871, the year Rogers's "Improvement in distilling naphtha and other hydrocarbon liquids" patent was allowed, kerosene prices were tumbling. At the same time, competition among the three railroads that carried crude from the Oil Regions to refineries in Cleveland, Pittsburgh, Philadelphia, and New York, drove down the profits of those companies. Tom Scott, the head of the Pennsylvania Railroad conceived of the "South Improvement Company" as a means of restoring the distressed profits of the two industries, oil refining and railroad transport, with particular attention to the profitability of those refiners and railroads that invested in the South Improvement Company. The Company included three mutually hostile railroads, the Pennsylvania, the New York Central, and the Erie, and a select group

---

[10] For my understanding of the mysterious South Improvement Company, I am indebted to Chernow's biography of John D. Rockefeller, Sr. [8].

of petroleum refiners. In the early part of the SIC's short history, the Standard Oil Company of Cleveland chose to associate itself with the South Improvement Company, while refiners in the Oil Regions and those in New York were excluded from membership in the SIC.

Scott's dark plan was this: the railroads would partition the oil trade, each railroad company carrying a fixed percentage of the total work. The prices of oil transport would be raised, but members of the SIC would benefit by preferred pricing and would receive "drawbacks" on transportation fees paid to the SIC-affiliated railroad companies by their rivals, the non-SIC refiners.

The SIC in was in place in February 1872: prices on transport of oil doubled for everyone except a few refiners in Cleveland, Pittsburgh, and Philadelphia. In Oil Creek, the newly formed Petroleum Producers' Union agreed to sell petroleum to Oil Creek refiners only. The reaction of the Oil Creek producers was vehement and sustained. By March, the Standard Oil Company and its president John Rockefeller had been targeted by the Petroleum Producers' Union; ninety percent of the employees of Standard Oil were temporarily laid off. Maintaining their plan to isolate the refiners in the Oil Regions but realizing their error in excluding the New York refiners, members of the South Improvement Company began courting the New York refiners. The SIC partners were too late: the New York refiners sided with the Oil Creek Refiners in opposition to the South Improvement Company.

As the sinister intent of the South Improvement Company became evident and understood, the SIC agreed to meet with the New York petroleum refiners. Henry Rogers, 32 years of age, a veteran of refining operations at the Oil Regions, partner in the celebrated Charles Pratt Refinery, and refining technology patentee, represented the New York refiners in negotiations with Tom Scott, the architect of the SIC. Rogers's confrontation with the railroads and the South Improvement Company took place in New York on March 25. The railroads, acknowledging the unfairness of their conspiracy, caved. At the meeting's conclusion, the railroads agreed to cancel their scheme of preferred pricing for SIC members.

As events became known, John Rockefeller was widely perceived to be the evil force behind the SIC, although the SIC was an invention of the railroad companies. Rockefeller, whose name had been not well known before the SIC venture, was suddenly both known and frequently vilified. Rockefeller had been denied access to the SIC-NYC refiners meeting. In reality, Standard Oil had significant presence in New York, but the company was not known to have refining operations in New York. In acquiring various refineries in New York, Standard Oil had new affiliates operate under their former identities. Rockefeller would indeed take interest in the SIC-New York refiners meeting.

Despite the failure of the South Improvement Company and the shade it cast on Rockefeller and Standard Oil during five weeks in early 1872, word of Standard Oil's cash reserve, its technological skill, and its fearsome business practices had a fright-inducing effect on Standard's competitors. As the SIC drama played, Standard Oil was able to acquire the refineries of several of the company's strongest competitors.

Whatever the unkind thoughts of Standard Oil's critics and competitors, the company's skill and strong finances seem to have inspired in its competitor refineries a mixture of terror and admiration. Financially, the most attractive outcome for the company's competitors was an offer to be purchased by Standard Oil, especially if payment were in the form of equity in Standard Oil. The less happy consequence for a refinery that had drawn Standard Oil's attention would be some form of refinery-annihilation.

During and after the short drama of the South Improvement Company, which appears to have served as a personnel recruiting event for Standard Oil, Standard Oil's consolidation of the petroleum refineries continued. William Rockefeller had opened an office of the Standard Oil company in New York in late 1860s, well before Standard Oil had any real operations there. New York, it appears, was an early apple of Rockefeller eyes. Not long after the drama of the South Improvement Company, several changes in ownership of petroleum refining assets in New York were announced or became evident. Standard Oil had acquired several refineries along Newtown Creek before it un-secretly courted the celebrated Pratt Oil Works.

When Standard Oil first indicated its interest in the Pratt Company and Astral Oil, it was Rogers who argued in favor of merging with Standard Oil while Pratt resisted. When the Pratt partners finally agreed on terms with Standard Oil, both took shares in Standard Oil in lieu of cash, a decision that made both Pratt and Rogers even wealthier than they had been. Within the management of Standard Oil, both Rogers and Pratt assumed new and broad responsibilities.[11]

Charles Pratt held special standing in New York's petroleum community. From the time of William Rockefeller's earliest presence in New York, even before the Standard Oil Company was chartered, the Rockefellers courted New York's petroleum community and established a discrete network of petroleum specialists. The company relied on its New York network as it restructured the huge and profitable organization. In 1883, as the Brooklyn Bridge was readied for its first traffic, Charles Pratt and four fellow New Yorkers chartered the Standard Oil Company of New York.

Rogers knew and understood more about refining technology than did Pratt or Rockefeller, and, as he assumed an office in the company's lower Broadway headquarters (Standard Oil moved its offices from Cleveland to New York in 1883), Rogers demonstrated previously untapped ability in management. With the Pratt company, he had developed methods for isolating the several naphtha components, including gasoline, which became of increasing commercial significance in succeeding decades. At Standard Oil, his work included acquisition of crude oil from producers, management of transfers of crude oil by pipeline from production site to refinery, and, eventually, for all manufacturing operations of the company.

In the period following Standard Oil's acquisition of the Pratt Refinery and Astral Oil, Charles Pratt devoted his effort to building the Pratt Institute, which opened in 1887. Pratt died in 1891; Rogers became a VP of Standard Oil in 1890; Pratt's eldest son Charles became Secretary of Standard Oil.

---

[11] Charles Pratt's Oil Works may have had an affiliation with Standard Oil that predated sale of the Pratt businesses to Standard Oil. An 1873 map of Long Island City identifies the former Long Island Oil Refinery at East River and West 10th Street in the Hunter's Point section of Long Island City as property of Standard Oil Company. Two lots that adjoin the Standard Oil property are identified as "Cha's Pratt's Oil Works" [9].

References

1. Leary L (ed) (1969) Mark Twain's correspondence with Henry Huttleston Rogers 1893–1909. University of California Press, Berkeley and Los Angeles
2. Charles Pratt, an Interpretation 1830—1930 Brooklyn, privately printed; foreword by Pratt F, collection of the Rockefeller Archive Center, Sleepy Hollow, New York
3. Lionel Pincus and Princess Firyal Map Division, The New York Public Library. (1868). https://digitalcollections.nypl.org/items/64b498f8-4a89-c936-e040-e00a1806393b
4. This information is constructed of information disclosed in corporations' chartering documents found at the New York County Clerk's Old Records office, 31 Chamber St, NYC. The documents indicate the establishment of the Devoe and Pratt Manufacturing Company in 1868; principals of the company included Devoe, Pratt, and Raynolds. The company documents indicate Devoe and Pratt organized as a can manufacturing and oil retailing business, an area of interest common to the two, while the Charles Pratt Oil Works continued independently to build a position in petroleum refining
5. Rogers H (1871) Improvement in Distilling Naphtha and Other Hydrocarbon Liquids, US Patent 120539, October 31, 1871
6. Anon (1872) Petroleum and its products. Sci Am supplement (May 18, 1872) 342
7. Williamson J, Daum A (1959) The American petroleum industry, 1859–1899 the age of illumination. Northwestern University Press, Evanston, p 279
8. Chernow R (1998) Titan, the life of John D. Vintage Books, New York, Rockefeller, Sr
9. Lionel Pincus and Princess Firyal Map Division, The New York Public Library. (1873). *Hunters Point. Part of Long Island City.* Retrieved from https://digitalcollections.nypl.org/items/510d47e2-634e-a3d9-e040-e00a18064a99

Chapter 8
Acid and Copper: The 50-Year Partnership of John Brown Francis Herreshoff and William Nichols

That a young man, armed with ambition and a fresh university degree, would turn his back on Manhattan to accept an entry level job at a chemicals factory on Newtown Creek is, in the present day, almost inconceivable, but in the late decades of the nineteenth century, Newtown Creek attracted talent. The rough-edged district was in process of becoming a center for innovation in the production of refined materials, from sugar to manufactured fertilizers and high-energy fuels. A handful of manufacturers of reagent chemicals supported the larger industries. Every valuable thing prepared at Newtown Creek gained value through processes that required reagent chemicals. Reagent chemicals convert the raw to the refined. Each day at Newtown Creek, tons of crude materials were crushed or fractionated, treated with reagent chemicals, distilled and re-distilled, or crystallized and recrystallized. Barges of coal, shiploads of raw sugar, barrels of crude oil, and wharves-full of dead horses were treated with reagents to form patent leathers, clean-burning lamp oils, glues, glass, rum, fabric dyes, phosphate-rich fertilizers, or crystalline white sugar. Barges of rough and crude sailed into the Creek, and barges of refined and expensive sailed out. Newtown Creek was an ideal place for a chemistry-minded college graduate to seek new fortunes.

William Nichols was 18 years old with a university training in chemistry when he began a career at the Laurel Hill Chemical Works, midway along Newtown Creek, on its northern shore.

8.1 Two Young Men Attend University, One in New England and One in New York

William Nichols was born in Brooklyn in 1852 and began his university education at Brooklyn Polytechnic Institute (now a campus of New York University). He completed his degree at the University of New York, which was later renamed New

P. Spellane, *Chemical and Petroleum Industries at Newtown Creek*, History of Chemistry, https://doi.org/10.1007/978-3-031-09629-7_8

York University.[1] At NYU, Nichols became a protégé of the esteemed chemistry Professor Draper before graduating in 1870[2] [1].

As William Nichols was diligent in his pursuit of chemistry and a university degree, John Brown Francis Herreshoff was tentative, skeptical of the importance of a university degree. Just as Nichols arrived at the University of New York, Herreshoff registered at Brown University in Providence. In July 1867, in the company of his father (a graduate of Brown), Herreshoff traveled from his home in Bristol, Rhode Island to Providence to register as a member of the Brown University class of 1871. One can imagine the younger Herreshoff signing his name in the class ledger, turning the ledger back to the Registrar, anticipating the interrogation that would follow. One imagines a discussion and the Registrar's pause before writing "select" in the space for "Degree." Fifty-two students, the class of 1871, registered that and the following day. Forty-nine would sign on for degrees identified as A.B. or B.P.,[3] only Herreshoff and two others came to Brown for "select" classes. Herreshoff's great-grandfather was John Brown, a banker and merchant, a man of wealth, prominence, and some notoriety in Providence, and a founder of Brown University. The 17-year-old Herreshoff gave no indication of entitlement or sentimentality for family tradition at Brown; his choices indicate what he sought at Brown. Herreshoff took classes in Chemistry, Physics, Mechanics, Rhetoric, and Trigonometry and worked in the University chemistry lab. His career suggests that Herreshoff acquired at Brown an understanding of the fundamental science and practices of chemistry and chemical analysis. Herreshoff's time at Brown may have sharpened an innate aptitude for engineering and construction that appears to be a genetic asset of the Herreshoff family.[4]

---

[1] A few years after the University of New York began operations in 1831, the College of the City of New York began receiving students. The former would become New York University, while the other became City College of New York. When CCNY was joined by additional campuses, the collection of colleges came to be called the City University of New York (CUNY). Late in the nineteenth century, NYU relocated its undergraduate campus from Greenwich Village to a new campus at University Heights in the Bronx. In the 1980s, NYU sold its University Heights campus to CUNY; the campus, largely unchanged, is now home to Bronx Community College, a campus of CUNY. Although he was there for no more than a few years, William Nichols provided financial support to NYU; among other gifts, he provided funds for Nichols Hall, the present-day chemistry lab at Bronx Community College.

[2] John William Draper, born near Liverpool in 1811, had studied chemistry at the University of London, moved to the US soon after earning his degree [1]. Draper studied medicine at the University of Pennsylvania, taking an interest in physiology that he maintained through his career. Physiology inspired Draper to examine the chemical and mechanical effects of light incident on chemical substances. With the support of NYU's president Frelinghuysen, Draper built a medical department at the University. Interest in light and chemicals (photochemistry) led Draper to photography. Draper is credited with taking the first photographic portrait of a human face, his sister's, made at a studio at Washington Square. The patented techniques of Daguerre had been made available by the French government; Draper, armed with theories concerning the photosensitivity of particular compounds, engineered his photographic plates for optimum responsiveness to studio lighting.

[3] A.B. and B.P. refer to Bachelor of Arts and (most likely) Bachelor of Philosophy, information provided courtesy of the Brown University Library staff.

[4] Two of Herreshoff's brothers formed the Herreshoff Manufacturing Company in Bristol. The company designed and manufactured power boats and sailing yachts. His brother Nathanael Greene

In the course of the sustained professional career that followed his short tenure at Brown, Herreshoff elevated the scale and sophistication of industrial chemistry in the United States.

In four years in Providence, Herreshoff completed eight classes and borrowed a single book (on Human Physiology) from the Brown library. His grades in Chemistry were unexceptional, but his laboratory skills were extraordinary: on the recommendation of Brown's one professor of Chemistry, the University hired the degree-less twenty-one-year-old a member of staff with title Assistant in Analytical Chemistry. He stayed for just a year. In 1872, Herreshoff left Brown to find a place in New York's new chemistry and materials economy. Employment came easily, first with Dr. Charles A. Seely, later with the Silver Spring Dyeing Establishment, and later still, with the analytical chemist William M. Habirshaw [2].

The career William Nichols began at Laurel Hill in 1870 lasted a lifetime. The 18-year-old Nichols arrived at this first job with two advantages: (1) an impressive education in chemistry and (2) the friendship of Charles Walter, one of two founders of the Laurel Hill Works. Within a few years, circumstances at Laurel Hill changed: Walter's partner August Baumgarten retired in 1871; not long after that, presumably to Nichols's great distress, Charles Walter died in a swimming or sailing accident in England. Following Baumgarten's retirement, Nichols's father, G. H. Nichols (Nichols's third advantage) had invested in and assumed financial responsibility for the company. By 1875, the younger Nichols found himself the director of operations at Laurel Hill company, owned by his father and rechristened "G. H. Nichols and Company."

William Nichols and Francis Herreshoff were contemporaries. It is unclear how Nichols and Herreshoff first crossed paths but unsurprising that they did. They were of similar age, background, and education, and they had similar professional ambitions. What did the two twenty-somethings talk about when they first met? Who initiated the contact? Was there a coffee house where there is now a Starbuck's at Astor Place? The chemical societies and professional meetings did not yet exist.[5] Their conversations, their brainstorming, would not be unlike the conversations of ambitious 26-year-olds of this generation: how they could do what others did but do it faster, smarter, and better. Nichols and Herreshoff must have spoken of sulfuric acid, the *sine-qua-non* of industrial chemistry and the Nichols company's main product. The manufacture of this harsh, corrosive liquid, which, although it continues to be manufactured and used in greater volume than any other substance, may be the least chic product in an unglamorous profession. But, in the second half of the nineteenth century, as the country began to flex its industrial muscles and realize its ability in production of chemicals and materials, sulfuric acid would be fascinating to ambitious chemists, even if not to the whole population.

Herreshoff was a naval architect, the designer of a series of America's Cup defending yachts in the late nineteenth and early twentieth centuries.

[5] Nichols and Herreshoff would help found the New York Chemical Society, the group that grew to become the American Chemical Society.

In 1876, the 24-year-old Nichols hired the 26-year-old Herreshoff, appointing him "Superintendent of Operations" at the Nichols company. Nichols and Herreshoff were at the helm of a 10-year old company, that had not yet achieved any real momentum. In production of sulfuric acid, the Laurel Hill Works would be a distant second to the nearby Kalbfleisch Works.

For two young entrepreneurs, control of production at the Laurel Hill plant, located at the center of Newtown Creek, ground zero of New York's new chemicals and petroleum industries, New York's busiest industrial zone, in the harbor that had become the country's center for trade and export, there was no better opportunity. Production of sulfuric acid placed the two young men at the center of the country's new economy of fuel and materials. There was no better opportunity, but there was a major and immediate challenge: the other sulfuric acid producer at Newtown Creek.

When William Nichols and Francis Herreshoff began their work at Laurel Hill, the premier manufacturer of sulfuric acid in New York was Martin Kalbfleisch, located about a quarter mile upstream of Laurel Hill. A contemporary historian effused in his description of the Kalbfleisch works [3]:

> Martin Kalbfleisch and the "Bushwick Chemicals Works...are among the most important and extensive Chemical manufactories in the United States.... One of the chambers, for manufacturing Sulphuric Acid, is two hundred seventeen feet long by fifty feet wide, no doubt the largest in existence, and is a model in every particular. Among the noticeable objects that attract the attention of visitors, are three Platina Stills, imported from France, at a cost of about fifteen thousand dollars each. ... Of Sulphuric Acid they have a capacity for producing three hundred thousand pounds weekly.

If the Kalbfleisch company was among the country's most important producers of sulfuric acid, as the historian suggests, the company's likely source for sulphur was brimstone, elemental sulfur, and their method the chamber process. The Platina Stills that the historian raves would enable the Kalbfleisch plant to concentrate the chamber acid. Brimstone was more expensive than other sulfur sources, certainly more expensive than pyrite ore, and was thought to produce the cleanest sulfuric acid. If the Nichols company sought a competitive advantage in acid production, it would have to be by an alternative and innovative method.

When Nichols senior acquired the Laurel Hill Chemical Works (Fig. 8.1), the site's products were sulfuric and other mineral acids that can be prepared in the treatment of various minerals with sulfuric acid. In beginning his work at the Nichols company, Herreshoff seemed confident of a long tenure with Nichols: he marked out new areas of interest, chemical processes, and products that would define the company's early years and, over time, generate great wealth for the company and lay a basis for its expansion. Herreshoff's first patent, issued seven years after he joined the company, did not concern acid or chemicals. Herreshoff's patent did suggest that the reinvigorated company had interests in new materials and new processes. Herreshoff's 1883 patent described a water-jacketed furnace for roasting copper-rich mineral ores.

There must have been questions asked when the Superintendent of Operations announced to the owners of G. H. Nichols his first patent. Copper ores? Did Herreshoff know that Nichols was not a mining company? He did. It becomes clear

Fig. 8.1 The Laurel Hill Chemical Works 1881, G.H. Nichols and Co Proprietors [4]

to a reviewer of his career that Herreshoff, and presumably Nichols junior, had particular insight and understanding of the chemistries of copper and sulfur and the affinity of each element for the other. Herreshoff's long and successful career, documented by a long string of patents, suggests that he and Nichols were in process of building companies that produced high-value materials, chemical compositions that are central to advanced manufacturing societies. Herreshoff and Nichols worked together for fifty years, capturing and reworking sulfur (for its role in sulfuric acid) and copper (for its role in materials of construction and electronics).

The central importance of sulfuric acid to the chemicals industry and, through chemicals, to industrial economies is, to most people, unappreciated. Sulfuric acid is not a household item; it is dense and fiercely corrosive. Within the chemicals industry, sulfuric acid is used to make other chemicals, the starting materials or processing reagents that comprise or treat or make ready for end use nearly every manufactured material. In the larger economy, sulfuric acid's largest applications are the manufacture of fertilizer and the refining of petroleum.

The central importance of copper to civilization may be more widely appreciated than is that of sulfuric acid. We handle copper, in jewelry, in coins, in materials of construction, and know that bronze, an alloy of copper and tin, enabled the Bronze Age. But in the age of the Edison companies and continuing in the age of microelectronics, high purity copper achieved new applications. Herreshoff's first patent, and most of the 27 that followed, concerned the preparation or refining or improvement of handling of oxides of sulfur, sulfuric acid, and high-purity copper. In choosing to pursue the preparation of sulfuric acid and refining of high purity copper, Herreshoff and Nichols revealed impressive ambition. In time, their achievements would match their ambition.

8.2   A Brief History of Sulfuric Acid

Large-scale industrial production of sulfuric acid began in England in 1733. Joshua Ward developed a two two-step "chamber process" that made possible, for the first time, high-volume production of sulfuric acid. Elemental sulfur ("brimstone") reacts with potassium nitrate ("niter") to form sulfur trioxide; sulfur trioxide reacts with water to form sulfuric acid. A few years later, John Roebuck scaled-up production by carrying out the reactions in lead-lined chambers. Sulfuric acid was more commonly identified as oil of vitriol in its early industrial years.

Industrial production of sulfuric acid began in the United States only after the constitution was signed, and indeed, not far from Constitution Hall. In 1793 John Harrison set up a chamber process production site in Philadelphia.[6] The European chemistry powerhouses had initiated industrial production of sulfuric acid 50 years before the US; the US makers benefitted by the Europeans' ingenuity, the teachings of their patents, and the plant equipment that they manufactured and sold.

Industrial production of sulfuric acid in New York began three decades after production began in Philadelphia: in 1823, the New York State legislature chartered "The New York Chemical Manufacturing Company." The new manufacturing company advertised its products: "blue vitriol, alum, oil of vitriol, aquaforte, refined camphor, saltpeter, borax, copperas, corrosive sublimate, calomel, and other drugs, medicines, paints, and dyers' colors." In its earliest charter, the Manufacturing Company was enjoined from participating in any banking activities, but this limitation was eased a year later when the Legislature allowed that a minor fraction of the company's assets be directed to accepting deposits and lending. The company succeeded in its primary mission, manufacturing and selling chemicals, but, as it grew, the directors of the company, bankers in training, took greater interest in the company's minor activity and expanded it. To supervise the chemicals manufacturing, the company employed a young Dutch immigrant who did a masterful job of it, Martin Kalbfleisch.[7]

Sulfuric acid production was improved again in 1835 by Joseph Gay-Lussac who developed an oxidation method that did not require nitre. Sulfur dioxide, formed easily in the roasting of brimstone or pyrite mineral, reacts with nitrous acid (formed

---

[6] John Harrison had been a student of English chemist Joseph Priestley who had met Benjamin Franklin in London in 1766. Priestley and Franklin were fellow "electricians," early investigators of electricity. When Priestley was forced to leave England in 1794, Franklin helped Priestley settle in the United States.

[7] See description of The New York Chemical Manufacturing Company in Chap. 3 of this book.

in the reaction of nitrous oxide with oxygen in water) to form sulfuric acid. The nitrous oxide could be recycled to form more nitrous acid. Glover towers serve to recover the oxides of nitrogen and cool the highly exothermic reaction. Water and "chamber acid" are sprayed as a mist onto the hot oxides of sulfur and nitrogen, and the acid forming reaction proceeds. Water can be distilled from the "chamber acid" to form more highly concentrated sulfuric acid.

8.3 Sulfuric Acid and Copper Often Have a Common Mineral Source[8]

The raw material needed for production of high-quality copper is simple: a copper-rich mineral ore. The raw materials from which one can prepare sulfuric acid are nearly as simple: sulfur or a sulfur-rich mineral, oxygen, and water. All three are abundant; the latter two are almost without cost. Copper metal can be recovered from ore by "reduction," the conversion of copper oxide or copper sulfide to copper metal. For its use in sulfuric acid, the sulfur found in nature, either as elemental sulfur or in metal-sulfide compounds, must be "oxidized" (combined in a chemical reaction with oxygen). There are two common oxides of sulfur: sulfur dioxide and sulfur trioxide. Sulfur trioxide, also called sulfuric anhydride, combines with water to form the acid. The hard part in making good copper or good sulfuric acid is the "refining," processing the raw materials to render the material of interest, copper or sulfur, essentially free of impurities. To prepare high purity samples of materials in large volume and at an acceptable cost, an engineer must design reactors and reaction conditions that enable the chemical processes to proceed and do not damage the reactors in which the chemical processes take place.

Except for the most noble of the lot, metals are found in nature as compounds with oxygen (metal oxides) or with sulfur (metal sulfides), and the oxides and sulfides of metals bear little semblance to the metals. While small amounts of gold, silver, and copper can be found as metals, our long-ago ancestors worked out methods for extracting copper, iron, and tin metals from their minerals. Smelting describes a set of rudimentary chemical processes that enable recovery of useful metals from accessible and sometimes abundant minerals. If an iron-rich mineral ore is heated to high temperature in the presence of coal, coal can give up electrons which the iron mineral can accept: the coal is said to be oxidized as the ore-sourced iron is reduced. Pyrite ores comprise great amounts of iron and sulfur and are distributed throughout much of the globe [6].

Iron copper sulfide minerals (chalcopyrites) can serve as sources of copper, iron, and sulfur. Roasting chalcopyrite ores (which comprise roughly equal masses of iron, copper, and sulfur) can enable the recovery of copper and iron as metal oxides

---

[8] I am indebted to Curtis Craven's book *Copper on the Creek* [5] which inspired me to learn about copper manufacture at the Nichols companies which, in turn, led me to examine JBF Herreshoff's history of invention.

(cuprous oxide, cupric oxide, ferrous oxide, and ferric oxide) and sulfur in the form of oxides, including sulfur dioxide and sulfur trioxide. The sulfur oxides can be further oxidized then combined with water to form sulfuric acid.

In a second chemical process, the metal oxides may be converted to metals in a "reduction" reaction: a readily oxidized material such as coke or charcoal (or the sulfide ions present in the chalcopyrite mineral) provides electrons that "reduce" the metal oxide to metal and releases the oxygen as carbon dioxide or sulfur dioxide.

As the new industrial economies demanded greater quantities of high-quality sulfuric acid, the price of brimstone rose. At some moment in the second half of the nineteenth century, early in this country's industrial period, the notion of recovering sulfur from sulfide minerals, made economic sense. The further idea, recovering both metal and sulfur from a metal sulfide, making use of the noxious waste gas produced at smelters, made even better sense. A producer that developed methods for production of both high-quality sulfuric acid and high purity copper from a mineral source could enjoy two revenue streams flowing from a single mineral source.

If his patent record reflects the chronology of Herreshoff's work at G. H. Nichols, the new employee's first project concerned a technology that appeared to support production of copper, in which business the Laurel Hill company had no commercial interest. But if Nichols's and Herreshoff's strategy for succeeding in the business that Martin Kalbfleisch dominated involved developing two revenue streams from one raw material, they would have been both bold and wise.

8.4   Herreshoff's Patent Portfolio: A Sustained Interest in Both Copper and Oxides of Sulfur[9]

The Nichols Chemical Company, and the General Chemical company that succeeded it, and the Nichols Copper Company (Figs. 8.2 and 8.3), spun off from Nichols Chemical at the time of the merger of acid producers, benefitted from the continued improvements in methods of production and design of production machinery that Francis Herreshoff generated in his tenure at the several Nichols companies. Herreshoff is the sole inventor on twenty-eight US patents assigned to Nichols or the various Nichols companies; all address production of materials: including processes, methods, or designs and maintenance of machinery of production. The Herreshoff patents identify and teach methods for the manufacture of sulfuric acid, including the design of ore-roasting furnaces, the production of sulfur dioxide in the roasting of sulfide ores, its recovery and purification from furnace gases, the optimization of the "contact process" for conversion of sulfur dioxide to sulfur trioxide, or methods for production of highly concentrated sulfuric acid in the addition of catalytically prepared sulfur trioxide to moderately concentrated sulfuric acid (Figs. 8.4 and 8.5). Herreshoff developed methods that enable production of highly concentrated acid in continuous, rather than batch, preparations.

---

[9] See Comment at end of this chapter.

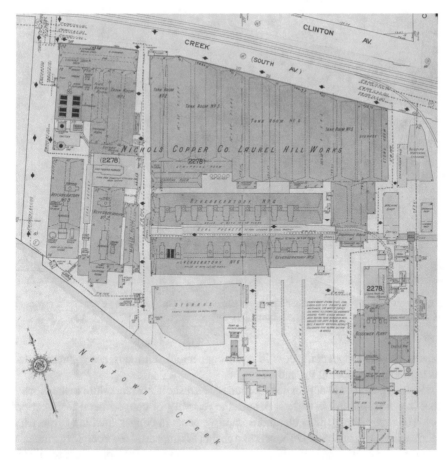

Fig. 8.2 The Nichols Copper Company site at Laurel Hill (Courtesy of The New York Public Library [7])

A number of the Herreshoff patents address the design of ore-roasting furnaces (Fig. 8.6) and emphasize the methods for efficient cooling of moving parts in roaster furnaces, minimizing their failure-rates, and optimizing their efficiencies. Herreshoff's earliest patent (1883) concerned improvements in the design of a copper smelting furnace. In smelting furnaces, the copper components of a copper-rich ore, such as the mixed metal "chalcopyrite," an iron-copper-sulfide, are reduced to a copper, iron, and sulfur composition that is enriched in copper (matte copper), as the sulfur components are oxidized to sulfur dioxide.

In a patent issued exactly 18 years (1901) after his first patent, Herreshoff described the design of a giant turn-table-like device for the continuous preparation of thin metal anodes (Fig. 8.7). Although the patent language does not specify an application, the production of copper anodes from molten blister copper could be of value in a large-scale process for electrolytic production of high-purity copper. In an electrolytic

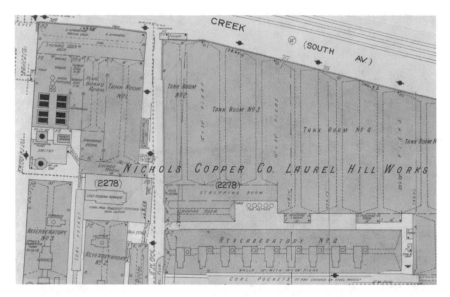

Fig. 8.3 Detail of Nichols Copper Company site at Laurel Hill, indicating electrolytic processes (Courtesy of The New York Public Library [7])

process, blister copper could be oxidized to copper ion at an anode, and re-precipitated as high-purity copper at the cathode.

Several patents that appear near the end of Herreshoff's career return to the process that he addressed in his earliest patent, the recovery of copper metal from copper ore. His first patent (1883) identified improvements in copper smelting furnaces; a Herreshoff later patent (1912) concerns a design for improved reverberatory furnaces, enabling preparation of high-quality copper matte, an ideal form of copper recovered from ore and ready for further refining in electrolytic processes.

8.5 Epilog 1: The Perkin Award

On a Friday evening in January 1908, the chemistry community in New York City met for dinner and the presentation of an award [12]. The evening's events brought together members of the Society of Industrial Chemistry, the American Chemical Society, the American Electrochemical Society, the Chemists' Club of New York City, and the Verein Deutscher Chemiker to witness presentation of the Perkin Medal to Mr. J. B. F. Herreshoff. The Society of Industrial Chemistry had established the Perkin Medal a year before; the first was awarded to its namesake, Sir William Henry

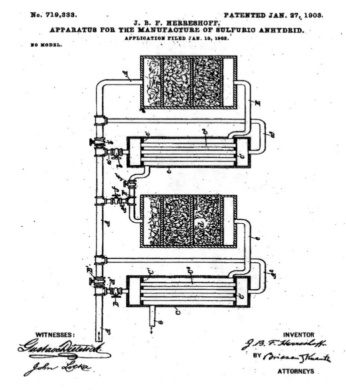

Fig. 8.4 Illustration from 1903 US Patent 719,333, Apparatus for the manufacture of sulfuric anhydrid [8]

Perkin, an Englishman.[10] Since that time the medal has been awarded to an American for excellence in industrial chemistry. In 1908, the first Perkin Medal awarded to a person not named Perkin was presented to J. B. F. Herreshoff.

Professor C. F. Chandler of Columbia University presented the medal "to our distinguished brother chemist, J. B. Francis Herreshoff" and spoke at length of Herreshoff's first thirty years of industrial work with William Nichols and the several Nichols companies [12]: "He has substituted scientific management for the old-fashioned rule-of-thumb in the conduct of all the operation which have come under his direction, and has thus developed the highest standard of factory control." Chandler describes Herreshoff's, and Nichols's, success with Herreshoff's first invention, the steel, water-jacketed furnace with a movable well, "now almost universally employed."

---

[10] William Henry Perkin studied the chemistry of aromatic compounds derived from coal tar, the so-called "coal tar dyes." He took particular interest in their applications in the vigorous dye and colorants industry.

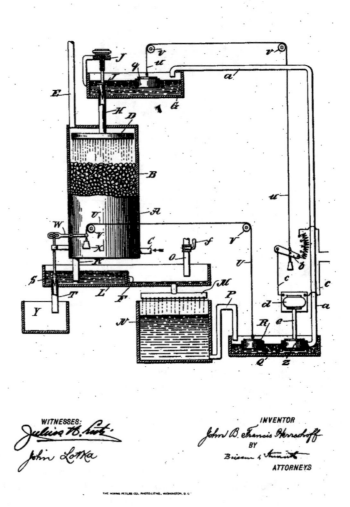

Fig. 8.5 Illustration from 1903 US Patent 737,625, Process of making sulfuric acid [9]

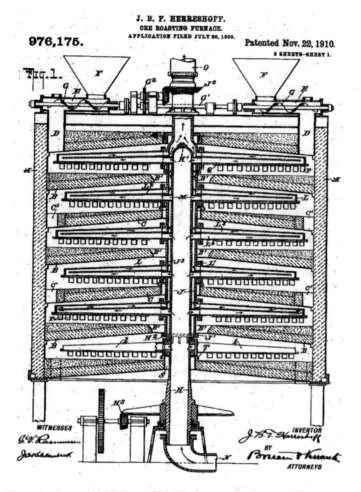

Fig. 8.6 Illustration from 1910 US Patent 796,175, Ore-roasting furnace [10]

Herreshoff's improvements to the chamber process for production of sulfuric acid reduced consumption of steam and derived greater economy in decreased consumption of niter. His "Herreshoff tower" improved denitration and concentration of chamber acid. Its steel plate construction and quartz lining rendered the Herreshoff towers "nearly indestructible." In maintaining high temperatures in the Herreshoff towers, denitration of the chamber gas is essentially complete and the chamber acid is of higher concentration. The acid can be further concentrated in distillation. Chandler claims that [12]:

> at one single establishment 100 tons of distilled sulphuric acid were produced daily from pyrites, that compared favorably with the concentrated acid made from brimstone, satisfying the demands at the time for purity, at a price at which brimstone could not profitably compete.

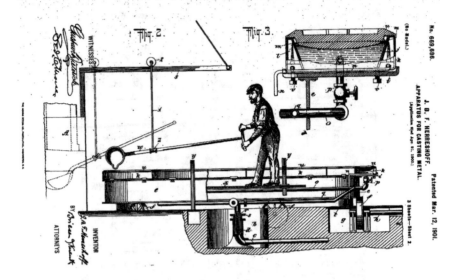

Fig. 8.7 Illustration from 1901 US Patent 669,696, Apparatus for casting metal [11]

On Herreshoff's design of roasting furnaces for pyrite fines: Chandler points to Herreshoff's 1896 patent that describes the furnace's removable stirrer arms. Herreshoff's series of patents on ore-roasting furnaces describe improvements to the standard "Gilchrist & Johnson" type furnace. A central shaft would turn sweeper arms over crushed minerals forcing the pieces of ore over circular furnace shelves, as liberated sulfur dioxide gas rises and is collected at the top of the furnace. Herreshoff designed removable, easily replaceable sweeper arms and, in later patents (several of which appeared after Herreshoff received the Perkin Medal), Herreshoff described designs that enable air-cooling of the central shaft and sweeper arms, designs that further extend the working lifetime of the furnaces.

Chandler points out that when the Herreshoff design was described, there were [12]:

> immense quantities of fines ores in various parts of the country that could not be sold on account of the inability to roast it, especially for making sulphuric acid, without great cost and trouble. This ore was used up very rapidly thereafter, without great cost or trouble. This ore was used up very rapidly thereafter, rendering available say 3,600 tons a day of fines pyrites which hitherto had been a kind of drug (sic) in the market. When one estimates that this means the production of more than 1,500 tons of sulphuric acid a day, he can see the importance of this furnace to that industry. … Up to this time, there have been sold 478 furnaces in the United States and 644 in Europe, making a total of 1122 furnaces.

Chandler calculates that, if all these furnaces were applied to the production of sulfuric acid and assuming each furnace to be capable of burning 3000 pounds of sulfur in 24 h, these 1122 furnaces would produce 2.5 million tons of sulfuric acid annually.

Chandler described Herreshoff's achievement with sulfuric acid production [12]:

It has made possible the production of sulphuric acid, of the highest strength, including anhydrous sulphuric acid, at a lower cost than by the old chamber system with the necessary concentrating plants in connection therewith, and now an acid virtually chemically pure can be supplied to the largest manufacturing industries in unlimited quantities at a lower price than previously known.

In electrolytic refining of copper, blister copper, the product of ore smelted in a reverberatory furnace, is cast to form thin long anodes, which, when immersed in an electrolytic cell and subject to an appropriate positive potential is oxidized to form copper ions. The copper ions generated at the anode migrate in the electrolytic cell to the counter electrode, the cathode, where, with great selectivity, copper ions are reduced to copper metal. The metal that is electrolytically deposited on the cathode is of very high purity. Before the Nichols company undertook electrolytic refining of copper, the method had been applied on limited scale at two sites, Chandler reports, but not well understood. Herreshoff addressed the work with characteristic focus and, in the first decade of the twentieth century constructed, on one part of the Laurel Hill site, what Chandler describes as the largest copper refinery in the world. The site's output was approximately one million pounds of copper a day, roughly one-fourth the entire world's output.

8.6 Epilog 2: Mergers and Acquisitions: The General Chemical Company[11]

It is likely that William Nichols followed Francis Herreshoff's work as closely as Herreshoff did and did so for strategic purposes. Herreshoff's patent estate, assigned to William Nichols or to the Nichols Chemical Company, created for Nichols a leading position in sulfuric acid production. William Nichols was alert to two trends in the chemicals industry in the last decade of the century: (1) advances being made by the German and English sulfuric acid producers in using "contact" processes to convert sulfur dioxide to sulfur trioxide, enabling them to prepare more concentrated sulfuric acid), and (2) the formation of large production companies, through mergers or acquisitions. Both trends, one technical and one organizational, threatened his company's strong position. Nichols proposed that, if twelve relatively small-scale producers shared resources in the production of fundamental chemicals, their collaborative effort could generate profits sufficient to enable growth in scale and variety of

---

[11] The General Chemical Company was formed in the merger of these twelve independent companies: Chappell Chemical Co. (production sites in Chicago and St Louis), Dundee Chemical Co. (Dundee, NJ), Fairfield Chemical Co. (Bridgeport, CT), Highlands Chemical Co. (Highlands, NY), Jas. Irwin & Co. (Pittsburgh, PA), Lodi Chemical Co. (Lodi, NJ), Martin Kalbfleisch Co. (Bayonne, NJ and Buffalo, NY), Jas. L. Morgan and Co. (Shadyside, NJ and Bridgeport, CT), National Chemical Co. (Cleveland, OH), Nichols Chemical Co. (Laurel Hill, Troy, and Syracuse, NY), Moro Phillips & Co. (Philadelphia, PA and Camden, NJ), and Passaic Chemical Co. (Passaic, NJ) [13, 14].

chemical product lines. Twelve companies working in collaboration, he suggested, could achieve more than they could in competition.

Nichols proposed a consolidation of a dozen small production companies to form a "General Chemical" company. Twelve companies, most privately owned, merged to form a corporation. In the words of the company historian [14]:

> The twelve concerns in question were found after careful appraisement to have an aggregate capital of $14,008,955 and it was believed, and as events have shown, correctly believed, that this aggregation would be large enough to realize most of those advantages, notwithstanding that the total capital of the country invested in chemicals was at that time about $238,000,000, and in heavy chemicals alone about $89,000,000.

> It may be interesting to dwell a moment on the details of the formation. On March 1, 1899, the various constituent concerns, after frequent conference and frank discussion, had settled upon a form of agreement providing for the transfer of all the properties to the newly formed corporation, full consideration to be paid to the sellers only after the values had been ascertained by careful appraisal. The basis of price was preferred stock to be issued for the tangible property and common stock for good-will and intangibles, although in some special cases common stock was issued for tangible property also...

> When the company was formed it took over from the constituents the manufacture of only fifteen different chemical products, but hardly any concern was making more than half a dozen and some not so many. The company is now [in 1919] making nearly one hundred distinct commercial products beside many hundreds of fine and chemically pure products and the list is increasing, and with each increase the risks of the business are diminishing.

The General Chemical Company describes in its 1919 review of its own history "the creation of a great coal tar products industry in American." General Chemical Company helped create the "National Aniline and Chemical Company." Until World War I, Germany dominated the production of dyes prepared from coal tar. The so-called "synthetic dyes" constituted a most significant part of chemical industry's portfolio. General Chemical and National Aniline would reunite as Allied Chemical and Dye Company in 1920, incorporating several other companies as well. Allied Chemical and Dye evolved into Allied Chemical (1981). Allied Chemical merged with the Signal Companies in 1985 to form AlliedSignal. AlliedSignal acquired Honeywell in 1999.

References

1. Archival material. The University Quarterly, Vol 1, No 3, New York University, July 1878, New York University Archives
2. Archival material. The Historical Catalogue of Brown University 1764–1934, published by Brown University, available at Google books, public domain, google-digitized
3. Bishop J (1868) A History of American Manufactures from 1608 to 1860, vol 3. Edward Young & Co, Philadelphia, p 196
4. Image of Laurel Hill works. Collection of the author
5. Cravens C (2000) Copper on the creek, reclaiming an industrial history. Place in History Press, New York
6. Rickard D (2015) Pyrite a natural history of fool's gold. Oxford University Press, New York

7. Lionel Pincus and Princess Firyal Map Division, The New York Public Library. (1884 - 1936). Queens V. 3, Plate No. 4 [Map bounded by Halle Ave., Clifton Ave., Newtown Creek, Hobson Ave. https://digitalcollections.nypl.org/items/90df2048-fd3d-05a8-e040-e00a18065fd4

8. Herreshoff JBF (1903) Apparatus for the Manufacture of Sulfuric Anhydrid. US Patent 719,333, January 27, 1903

9. Herreshoff JBF (1903) Process of making sulfuric acid. US Patent 737,625, September 1, 1903

10. Herreshoff JBF (1910) Ore-roasting furnace. US Patent 796,175, November 22, 1910

11. Herreshoff JBF (1901) Apparatus for casting metal. US Patent 669,696, March 12, 1901

12. American Chemical Society (1908) Proceeding of the American Chemical Society for the Year 1908. Eschenbach Printing Company, Easton, Pennsylvania, p 38

13. Haynes W (1983) American chemical industry, background and beginnings, in six volumes. Garland Publishing Inc., New York

14. Archival material. The General Chemical Company after Twenty Years: 1899–1919, published by the company in New York, 1919, pp 15–16

Chapter 9
The Standard Oil Company and New York City

> Besides her hoisted boats, an American whaler is outwardly distinguished by her try-works. She presents the curious anomaly of the most solid masonry joining with oak and hemp in constituting the completed ship. It is as if from the open field a brick-kiln were transported to her planks.
>
> Preface to Chapter XCVI, The Try-Works, *Moby-Dick, or the Whale* by Herman Melville, published by Harper and Brothers, New York, 1851.

The tryworks that Herman Melville describes had economic significance. American whaling ships were at once ocean-going fishing vessels and oil refineries. The practice of trying blubber cut from the corpses of freshly slain whales ensured the quality of the refined oil. The American whalers pursued whales with the finest fats and applied superior methods in refining and barreling their oil. Buyers at whale oil markets believed American oil to be of highest quality.[1]

The American whalers pursued the sperm whales for their especially fine oil. On capture of a sperm whale, the whaling crew would scoop spermaceti from animal's head case[2] and stripped the animal's flanks of thick sheets of blubber and their skeletons of fine whalebone [1].

Whaling was the first great American wealth-maker; American whalers established a new best-practice of a long-practiced endeavor. Working in a fierce natural world, building crews of sailors from all corners of the world, sharing the profits

---

[1] Trying blubber is like clarifying butter. In each case the as-received animal tissue is a combination of triglycerides (fats) and proteins. Trying involves heating the freshly cut "blanket strips" of blubber in copper cauldrons on the deck of the whaler. At high temperature, the triglycerides are stable, but the proteins decompose to a crisp solid. The fried protein pieces rise to the surface of the oil. Sailors tending the try-works could skim the deep-fried "crackle" from the oil and toss it to fuel the fire beneath the cauldron, or, as Melville describes, consume it. Whalemen, we understand, often developed a taste for crackle, just as kitchen cooks do for bacon left too long in a hot frying pan. Whale oil, skimmed of its decomposition-prone protein and cooled and barreled, would remain stable over a years-long voyage and, at market, command a good price.

[2] Melville describes the anatomy of the sperm whale's head and its case, "The Great Heidelburgh Tun," and sailor Tashtego's recovery of the spermaceti oil, "Cistern and Buckets," in chapters 77 and 78 of *Moby Dick* [1].

© The Author(s), under exclusive license to Springer Nature Switzerland AG 2022
P. Spellane, *Chemical and Petroleum Industries at Newtown Creek*,
History of Chemistry, https://doi.org/10.1007/978-3-031-09629-7_9

of successful ventures, whalers and whaling ships became legend, a descriptor of the early American experience. During the young country's whaling years, America lagged the European countries in industrial production and higher education, but the American whaling industry presented to the world a can-do approach to effort and production and established a standard of quality. Whaling and whale oils tied the young country to world markets. Before American coal or petroleum were mined and refined, whaling established an American position in the business of finding, refining, and marketing of premium fuels.

9.1   1854: The Invention of "Kerosene" and the Mining of "Rock Oil"

In the early-1850s, Abraham Gesner in New York and Luther and William Atwood and Samuel Downer in Massachusetts constructed refineries that produced fine lamp oils extracted from particular oil-rich coals. Coal was more plentiful and more easily had than were whales: oil from coal would cost less and serve a far larger market. In refining and marketing lamp oil extracted from coal, Gesner and his fellow coal refiners established a market for the refined oil that Gesner named "kerosene."

George Bissell was a lawyer in New York City who developed a fascination for "rock oil," as many others had and would, but Bissell was among the first to consider rock oil a possible source of kerosene. He had graduated from Dartmouth College in 1845; on a visit to his mother in Hanover, New Hampshire in 1854, Bissell called upon a friend at the College, Professor Dixi Crosby. Crosby had received a sample of rock oil from a former student. Crosby shared with Bissell what he knew of the oil and its source, the Oil Creek in western Pennsylvania.

Daniel Yergin writes of Bissell's thinking as he studied the rock oil sample [2]:

> Bissell knew that the viscous black liquid was flammable. Seeing the rock oil sample at Dartmouth, he conceived, in a flash, that it could be used not as a medicine but as an illuminant – and that it might well assuage the woes of his pocketbook.

Bissell[3] might have been aware of the patents recently awarded to Abraham Gesner, claiming production of an excellent lamp oil called kerosene, extracted from the oils of New Brunswick coal. He may have read an enthusiastic report in Scientific American describing a demonstration at the Art Union in New York of light produced in the combustion of two variations of Gesner's "kerosene" [4]. Bissell may have known of the company Gesner and others formed to refine and commercialize asphalt-based "hydro-carbon fluids" or been aware of the large refinery under construction at Newtown Creek.

Bissell persuaded his law partner J. G. Eveleth to evaluate the oil regions in Pennsylvania. Eveleth's assessment must have positive: Bissell, Eveleth, and five others chartered the Pennsylvania Rock Oil Company at the Office of the Clerk of

---

[3] The story of George Bissell's visit to Dixi Crosby is retold at the Dartmouth College website [3].

New York County in late December 1854. The world's first petroleum company would begin operations, the document indicates, on the first day of 1855. "The objects for which said Company is formed are to raise(?), procure, manufacture and sell, Rock Oil".[4]

Having established a company that would manufacture and sell rock oil, the Rock Oil partners sought to understand the chemistry of the substance they intended to manufacture and market: they sought the advice of a chemist. The company retained Benjamin Silliman, Jr., Professor of Applied Chemistry at Yale, provided him a sample of authentic rock oil, and requested a full chemical analysis. It is unclear how the Rock Oil company expressed its request or what the investors expected to learn, but Silliman's analysis was thorough, rich in detail, and unequivocal in its assessment. Silliman completed the task promptly.

The Rock Oil Company was less prompt in paying Silliman for his work.[5] On receiving his agreed compensation, Silliman provided his clients a concise and thorough evaluation of the crude petroleum [5].

In pursuing his analysis or rock oil, Silliman became the first petroleum refiner: he developed methods for fractionating petroleum, isolating on the basis of molecular weight the various components of petroleum's mixture of hydrocarbon compounds,[6] and established (and described in his report) methods for cracking petroleum (breaking larger, high molecular weight compounds into smaller, lower molecular weight, more volatile, and often more marketable compounds). He measured the quality and intensity of light generated in the combustion of the kerosene-like components of petroleum.[7] Silliman's detailed information and comments assured his Rock Oil clients of petroleum's commercial potential. His optimistic report inspired a new investment group in New Haven. Yergin adds, "Silliman himself took two hundred shares, adding further to the respectability of the enterprise, which became known as the Pennsylvania Rock Oil Company" [6].

Four years later, Edwin Drake, employed by a reorganized, recapitalized version of the Pennsylvania Rock Oil Company, drilled for oil as if he were drilling for water and succeeded. Yergin argues [7]:

> The essential insight of Bissell – and then of his fellow investors in the Pennsylvania Rock Oil Company – was to adapt the salt-boring technique directly to oil. Instead of digging for rock oil, they would drill for it.

Drake's was the first oilrig, the contrivance that freed the petroleum genie from its subterranean bottle. It would take a century and a half for the world to acknowledge the overwhelming environmental consequence of mechanized access to ancient pools

---

[4] The document chartering the Pennsylvania Rock Oil Company is filed at the Division of Old Records of the Office of the Clerk of New York County, at 31 Chambers Street, New York, NY.

[5] The story of the Rock Oil company's retaining Professor Silliman and its delay in paying for Silliman's work is presented at the opening of *The Prize* [2].

[6] Gesner had done a similar if less rigorous fractionation of coal oil.

[7] The founders of the New York Kerosene Oil refinery, in launching its kerosene-from-coal oil business, had sponsored a similar analysis and used the report's results to advertise the original kerosene.

of crude petroleum, but in the nineteenth century, Drake's discovery inspired a young nation. It convinced willing believers that petroleum and wealth could be had easily and in abundance. Silliman's report convinced investors that petroleum from the Oil Regions of western Pennsylvania could be refined to provide excellent illuminating oils.

Drake's rig inspired a new gold rush. Like the gold-diggers who had journeyed to California a decade before, roughnecks with ambition and gentlemen with money descended on the Oil Regions. Oilrigs soon covered the shores of the narrow creek, and before long crude petroleum soaked the area's fields and roads [8].

The new petroleum industry sorted itself into two groups: producers and refiners. The producers sucked oil from the ground, and the refiners adapted and improved methods pioneered by Gesner and Silliman for extracting and cleaning those parts of crude petroleum that met the demands of markets.

In the early years, production took place solely in the Oil Regions of Pennsylvania, while refining was done either near the oil fields or far from them, in cities (Cleveland, Baltimore, Philadelphia, New York) nearer to centers of marketing and distribution. Before long, a third business function took form: "transport" of crude oil from production sites to refineries and transport of kerosene from refineries to markets.

9.2 Cleveland and the Rockefeller Brothers

John Rockefeller and Maurice Clark began their careers in Cleveland, Ohio as clerks in the offices of commission merchants.[8] With no more than a few years' experience in the business of buying and selling grains, each believed himself capable of greater things. In 1859, the two formed a partnership. Each invested $2000 in the new venture; Clark was 32 years old and Rockefeller 20. "Clark &Rockefeller" advertised their partnership with confidence, declaring themselves ready to undertake "the management of any business." In the same year that Clark and Rockefeller launched their venture in grain-trading, Drake struck oil in Venango County, a hundred fifty miles east of Cleveland. News of the petroleum strike in Oil Creek stirred the imaginations of many business-minded individuals in Cleveland, but not those of Maurice Clark or John Rockefeller.

As refineries opened along the Cuyahoga River in Cleveland, Cleveland became an early leader in petroleum refining. It did so for good reasons: the city's proximity to the Oil Regions, its skilled workforce and established banking and investment communities, its abundant local coal and access, by both the Erie Canal and railways, to markets in eastern cities. As petroleum became an item of commerce in Cleveland, Clark and Rockefeller did take notice of petroleum and became acquainted with the oil-refining skills of Samuel Andrews. Chernow writes [11]:

---

[8] In understanding John D. Rockefeller's early interest in petroleum and the genesis of the Standard Oil Company, I relied on biographies by Allan Nevins [9] and Ron Chernow [10].

Clark and Rockefeller might have taken on consignment some of the first crude-oil shipments that reached Cleveland in early 1860, but it was the friendship of Maurice Clark and Samuel Andrews, an Englishman from Clark's hometown in Wiltshire, that drew Rockefeller into the business. A hearty, rubicund man with a broad face and genial manner, Andrews was a self-taught chemist, a born tinkerer, and an enterprising mechanic. Arriving in Cleveland in the late 1850s, he worked in a lard-oil refinery owned by yet another Englishman, C. A. Dean, and acquired extensive experience in making tallow, candles, and coal oil. Then in 1860, Dean got a ten-barrel shipment of Pennsylvania crude oil from which Andrews distilled the first oil-based kerosene manufactured in Cleveland. The secret of "cleansing" oil with sulfuric acid – what we now term refining—was then a high mystery, zealously guarded by a local priesthood of practical chemists, and many curious businessmen beat a path to Andrews's door.

Treatment of kerosene with sulfuric acid was not secret; the method appears in Gesner's patents describing preparation of kerosene distilled from coal oils [12].

The grain-marketers Clark and Rockefeller ventured into the new business of petroleum dealing, it seems, with reluctance, but the lure of the dollars changing hands in the petroleum trade must have been irresistible.

A small community of ambitious Clevelanders, comprising Rockefeller and Maurice Clark (grain traders), Clark's two brothers, and Sam Andrews (petroleum refiner) had shared interests in both grain trading and petroleum refining. Rockefeller and Clark sought to attach the group's diverse interests to the persons most interested in each. In a two-person auction for the petroleum interest, on a wintry afternoon in 1865, Rockefeller and Andrews secured the petroleum interests. Rockefeller (in colleague with Andrews) agreed to pay Clark $72,500 for the petroleum part; the Clarks took the grain interests [13].

As John Rockefeller secured a business position in the petroleum trade, his younger brother William began construction of his own Cleveland refinery. Within weeks of John's and Sam Andrews's acquiring a controlling interest in the Excelsior Refinery, Andrews and the two Rockefeller brothers organized their several holdings into two new companies. The first, Rockefeller & Andrews, consolidated the triumvirate's refining capacity in Cleveland by adding William's new "Standard Works" to the older Excelsior refinery. The second, Rockefeller & Company, established the refining company's presence in New York City [14].

This is the remarkable step. One wonders, what inspired the Rockefellers and Andrews to open an office in New York? Nevins provides a logical explanation that defends William Rockefeller's work in New York,[9] but their establishing a presence

---

[9] We could not find records of Rockefeller and Company's setting up office in New York at the Division of Old Records at the New York County Clerk's Office. Nevins reports [9], pp 40–41: "It is impossible to say exactly when just when William reached New York. But we know that notice of his withdrawal from Hughes Davis & Rockefeller appeared in the Cleveland Herald of September 18, 1865, that during the summer and fall the new Standard Works of William Rockefeller & Co., were going up, and that this refinery began operations about December 1. Presumably he was in New York early in 1866. Here contacts had to be established with buyers in New England, New York, and Europe, prices watched, and arrangements made for wharfage, warehouse-room, lighters, and the repairing and refilling of damaged barrels. William worked vigorously, became acquainted with the oil dealers and the city's bankers – for money could be borrowed at lower cost in New York than in Cleveland."

in New York suggests that Andrews and the Rockefellers foresaw an international market for refined petroleum. There was precedent for export of oils from New York: New York was a center for export of whale oil and coal-based kerosene. Still, the partners' opening an office at 181 Pearl Street in lower Manhattan suggests the extraordinary vision of the 20-something-year-olds. Andrews, the most experienced of the three, had not more than five years' experience with petroleum.

When William Rockefeller arrived in New York, the city, with its matchless harbor, had become the country's principal market for trading and export of cotton, whale and sperm oils, and other items of commerce. The Rockefeller-Andrews group set up a field operation not at petroleum's source but at its high value market. The three young petroleum refiners understood a reality established by the former owners of whaling ships: there is big money is in the sale of refined oils to European industrialists.

William Rockefeller would soon make the acquaintance of Josiah Macy, Jr. In New York.[10] The Macys, father Josiah and sons Charles A., William H., Josiah G. (Josiah, Jr.), Francis H., and John H., were descendants of a family of Nantucket whaling boat owners and sailors. Josiah Sr. had moved from Nantucket to New York in the early 1820s (he appears in New York City directories identified as "mariner.") Even as the Macys maintained positions in whaling (Josiah Macy & Sons continued to invest as partial owners of whaling ships through the 1860s), Josiah Macy and his sons established positions as merchants in lower Manhattan.

In the Doggett's New York Directory of 1847, the company is listed: "MACY, JOSIAH & SONS, commission and shipping merchants, 189 Front and manu-facturers of sperm oil and candles, agents for Freeman's sheathing and copper rollers, 266 South. h. 179 Broadway." A few years later (New York City Directory of 1855/56), the firm described itself, "MACY'S JOSIAH & SONS, commission merchants & manufacturers of sperm & whale oil & candles, 189 Front, 226 (as written but probably should be 266) South, and 525 Water. In the directories of the 1860s, the brothers are described by trade, "mer" (merchant), or "oils" and in still later year by title, "pres" (president) or "sec" (secretary).

9.3   The Rockefeller and Andrews Company Meets Flagler and Harkness

Henry Flagler arrived in Cleveland as William Rockefeller departed for New York. John Rockefeller and Henry Flagler had met when the two worked as grain-traders in Ohio during the Civil War. After winning and losing small fortunes, Flagler met and

---

[10] Reference to a letter co-signed by William Rockefeller and Josiah Macy, Jr.: Josiah Macy and William Rockefeller were cosigners "Petroleum – Memorial to Congress" published in Pittsburgh Weekly Gazette March 23, 1868) signed by producers and refiners of petroleum, arguing against the tax on refined petroleum, arguing that petroleum is serving mainly low-income consumers and presents a damaging burden on petroleum refiners struggling with excess-capacity in a depressed economy. Signatories are grouped by city (Cleveland signers include Henry Flagler). New York includes five: Josiah Macy, Jr., E. A. Wickes, Ambrose Snow, Herbert M. Warren, William Rockefeller.

married Mary Harkness, the niece of a prominent banker and businessman (who had experience working with oil refineries) in Monroeville, Ohio, Stephen V. Harkness [15].

When the elder Harkness moved to Cleveland, nephew-in-law Flagler followed. Flagler's reconnecting with Rockefeller had a profound effect on Rockefeller's new petroleum venture: Flagler and Rockefeller matched one another's ambition and drive in matters of commerce. Rockefeller admired Flagler, "full of vim and push." Rockefeller & Andrews reorganized itself in 1867 to form Rockefeller, Andrews & Flagler. Nevins [16] estimates that the new partnership received injections of new cash, estimated $50,000 from Flagler and $60 to $90,000 from Stephen Harkness. These investments supported additional refining capacity at the company's two Cleveland refineries, new company-owned tank cars, and new depots in the Oil Regions. In recollecting that moment, Rockefeller estimated that daily production of 500 barrels of refined oil in 1867 grew to 1500 barrels in 1869. With production and home office in Cleveland, the Rockefeller, Andrews & Flagler company maintained the New York office that William had opened at 181 Pearl Street.

In 1867, with JDR (John D. Rockefeller), Flagler, and Sam Andrews at work in Cleveland (doing business as Rockefeller, Andrews & Flagler) and WR (William Rockefeller) in New York (doing business as Rockefeller and Company), coordination among the companies foreshadowed the coordination of effort that would take fuller form in the firms' successor company. The Cleveland companies purchased and transported crude oil from the Oil Regions to Cleveland, refined it there, packaged it, and arranged for its transport to markets. The New York company received refined petroleum, in barrels or 5-gallon cans, from Cleveland and secured its further distribution to markets in New York, New England and, importantly, overseas.

9.4 The Standard Oil Company

In 1870, the prospects for the petroleum industry were uncertain. Among the variables were these: Cleveland was the dominant center for refining, and New York the largest petroleum market; prices for both crude and refined oils were unsteady; the railroad companies competed with one another with formidable ardor for the petroleum industry's business.

One thing that seemed certain is this: Rockefeller, Andrews & Flagler was the best managed of petroleum refiners and enjoyed exceptional access to investment capital. Believing their company and their plan to be the only hope for the unwieldy industry, Rockefeller, Andrews & Flagler reorganized their company one more time. The new organization could attract more capital with which it could acquire the resources of its strongest competitors. The five original partners recognized the opportunity and capital that new shareholders could provide but were determined to maintain control of their company. Rockefeller credited Flagler with a plan for achieving growth and maintaining control: the Rockefeller, Andrews & Flagler partnership would reorganize itself as a joint stock corporation [17].

Benjamin Brewster and Oliver Burr Jennings, both originally of Connecticut, had made money as merchants in the California gold-rush and carried their fortunes home to New York. Jennings married a sister of William Rockefeller's wife and had opportunity to become aware of Cleveland's best oil refinery. Recognizing a new gold rush and armed with capital and information from the earlier one, Brewster and Jennings sought to be part of the Cleveland company.

In the second week of January 1870, the owners of Rockefeller, Andrews & Flagler filed papers to create a new company in Ohio [18]:

> We John D. Rockefeller, Henry M. Flagler, Samuel Andrews, Stephen V. Harkness of Cleveland Cuyahoga County Ohio and William Rockefeller of the City, County, and State of New York have associated ourselves together... for the purpose of forming a body corporate for manufacturing Petroleum and its products under the corporate name of The Standard Oil Company.

The new company was organized in Cleveland under Ohio law, but New Yorkers were part of it from day one. William Rockefeller, having lived and worked three years in New York and marrying there, was a New Yorker. O. B. Jennings, while not a founder of the company, was the first "friends-and-family" investor. With fellow New Yorker William Rockefeller, Jennings was, in every sense, in the room as the company began operations.

A handwritten page and a half, signed and validated by the Clerk of the Court of Common Pleas of Cuyahoga County on January 10, 1870, established the Standard Oil Company. The simple charter indicates the company's purpose (the manufacture of petroleum and to deal in petroleum and its products), its capital stock ($1 million), the par value of a share of the company's stock ($100), and its venue, "the name of the place where said manufacturing establishment shall be located (illegible) for doing business is Cleveland City Cuyahoga County State of Ohio."

Eight weeks after its forming, the Standard Oil Company bought real property along the East River in New York harbor. With purchase of the Long Island Oil Company facility at Hunter's Point near Long Island City on March 9, 1870 for $35,000, Standard acquired a refinery in New York harbor, prime riverfront property, docks on the East River that could receive both rail-barge and ocean-going vessels, and easy access to the western terminus of the Long Island Railroad. The sellers of Long Island Oil Company, those signing the contract of sale, were Francis H. Macy and Josiah Macy, Jr. Josiah Macy was, by then, well known to William Rockefeller [19].

The young Standard Oil Company's earliest action, its first use of the funds invested by the five owners, was acquisition of a petroleum refinery with waterfront docks on the East River in New York harbor.

9.5 The South Improvement Company

Although John Rockefeller was widely blamed for the South Improvement Company's existence, he was not its originator. But Rockefeller appears to have embraced the controlled pricing model of the SIC and appreciated the company's planned hand-in-glove relationships among select refiners and owners of three railroad companies. Perhaps more importantly, the SIC gave Rockefeller opportunity to identify individuals with ambition and determination commensurate with his own. One could imagine John Rockefeller viewing the SIC as an employment fair of the most ambitious.

The South Improvement Company failed to achieve the self-serving intentions of its shareholders, but the company's presence had two significant results: (1) the SIC revealed to the industry and the wider public the vast ambition and ruthless manners of the most powerful players in the new petroleum economy, and (2), the private discussions among members of the SIC and with antagonists of the SIC brought a new group of individuals to the notice of John D. Rockefeller. The original shareholders of the South Improvement Company included one attached to a railroad (Peter Watson) and seven attached to refineries. Additional shareholders joined soon after; these included refiners from Cleveland (Henry Flagler, Oliver H. Payne, John D. Rockefeller) and from New York (Jabez Bostwick and William Rockefeller). Even as they appeared to be co-conspirators in the South Improvement Company, the five new shareholders may well have already agreed to work with the Standard Oil Company. Within a few years, all five would become officers of Standard Oil.

Among the vigorous opponents of the South Improvement Company were two of the most skilled refiners of petroleum: John D. Archbold from the Oil Regions and Henry H. Rogers, partner of Charles Pratt in Pratt's New York refinery. Within a few years of the SIC battle, Archbold, Pratt, and Rogers would also be officers of Standard Oil.

Jabez A. Bostwick, an early shareholder of the South Improvement Company, often identified as one of the few New York oil refiners to be part of the SIC, is likely to have been one of the business-savvy New Yorkers that William Rockefeller sought out. It is possible that Bostwick became an oil refiner only after he met William Rockefeller. Whatever his credential, it is clear that the Rockefellers developed a kinship with Bostwick. He soon became a part of Standard Oil's presence in New York.[11]

---

[11] In 1869 Bostwick was identified as an oil merchant in partnership with John Tilford. There is record of Bostwick's buying oil in the Oil Regions (crude or refined?), but it is difficult to determine the location of his New York refinery, if he owned one. Bostwick does appear to be a skilled capitalist and untimid man of affairs: he acquired, leased, sold, traded, sued, and was sued. He owned or directed companies ranging from petroleum to railroads. Shortly before his untimely death (attempting to save horses from a barn fire on his property in Mamaroneck, N.Y.), Bostwick had purchased a seat on the New York Stock Exchange. He may never have had petroleum under his fingernails, but he was key figure in Standard Oil's early years. John Rockefeller was among the first to call on the morning of Jabez Bostwick's death.

9.6   The Standard Oil Company's Presence Along Newtown Creek and the New York Harbor

The extent of Standard Oil's presence in New York City in 1872 is difficult to know with certainty. Even as companies sold out to Standard Oil, many continued to operate under their own, pre-Standard Oil, names. Standard Oil owned the former Long Island Oil Company at Hunter's Point on the East River and promptly expanded it holdings near there. The Bostwick company was likely to have been part of Standard Oil, and the Charles Pratt Refinery would soon be part of it (Fig. 9.1). Charles Austin Whiteshot prepared an encyclopedic description of the petroleum industry in 1905. In that tome [20], he lists 16 "Oil Refineries (w daily capacity in 55-gallon barrels) in New York in 1872".[12] Ten were located along or very near Newtown Creek. Several

---

We find no evidence of Bostwick's involvement with petroleum that predates William Rockefeller's arrival in New York. His earliest appearance in New York records (newspaper or city directories) is the 1866/67 Trow Directory of New York City, identified there as a merchant working at 42 Broadway and living at the St. Nicolas h(otel). Within a year, his office is at 76 Beaver Street, where it appears he made acquaintance of "broker" John B. Tilford, Jr., son of "banker" John B. Tilford, with office around the corner at 9 New Street and home at 297 Broadway. The 1868/69 directory identifies Bostwick & Tilford as "petroleum merchants" with offices at 66 Beaver Street. By 1869 and for several following years, Bostwick & Tilford "petroleum" have business offices at 60 Wall Street, possibly sharing space with John B. Tilford & Co Bankers. In 1871 and a few subsequent years, Bostwick and Tillford worked at 6 Hanover. In this period, Bostwick was identified in a New York Times article (June 2, 1870) for "non-payment of monies due in suit against money merchants in Wall Street for non-payment of a special tax on brokers and bankers. The matter was settled: New York Times, June 8, 1871: Jabez A. Bostwick vs. Joseph W. Wilday.

The Bostwick-Tilford partnership appears to be extinguished by 1873. It is likely that, by that time, Bostwick was aligned with Standard Oil. Tilford father and son may have been as well. That Bostwick became wealthy quickly is suggested by his holdings. In 1874, Bostwick kept office at 141 Pearl Street (Rockefeller's was at 181 Pearl), and he built a home at 9 East 64th Street. In the 1875/76 directory, Bostwick identifies his work at 141 Pearl and at the foot of West 66th Street. The latter site is identified on an 1877 map as Oil Works; the map makes clear the site included docks on the Hudson River and direct rail connection to the New York Central Railroad line that ran alongside the Hudson River, across the train bridge at Spuyten Duyvil and south to the rail yards at West 33rd Street.

In 1882, Bostwick was a signatory of the document with which Standard Oil of New York was established; he was named a trustee of the new corporation. Bostwick had another petroleum connection during that period: A. J. Pouch "petroleum" held offices at 60 Wall in 1870 (when Bostwick and Tilford did) and later at 6 Hanover (as did Bostwick and Tilford). In 1877 Albert J. Pouch is identified in the New York Times as partner with Jabez Bostwick (really working for Standard Oil) in a dispute concerning a leased oil refinery at Hunter's Point. A. J. Pouch appears in later directories as a "treas" at 110 Pearl Street and, after 1882, "oil" at 44 Broadway, addresses same for Jabez Bostwick, also shared by William Rockefeller.

[12] Charles Austin Whiteshot's List "Oil Refineries (w daily capacity in 55 gallon barrels) in New York in 1872.": 1. King's County Oil Works, Newtown Creek, Green Point LI; owners Sloan (Sone) and Fleming, 159 Front Street, N.Y.; 2. Pratt's Oil Works, Brooklyn E. D.; owner Charles Pratt & Co, 103 Fulton Street, N.Y.; 3. Empire Oil Works, Hunter's Point East River, Long Island City; owner R. W. Burke, 181 Pearl Street, N.Y.; 4. Queens Co. Oil Works, Newtown Creek, Long Island City; owner R. W. Burke, 181 Pearl Street, N.Y.; 5. Franklin Oil Works, Newtown Creek, Brooklyn E. D.; owner R. W. Burke, 181 Pearl Street, N.Y.; 6. Olophine Oil Company, Greenpoint Long

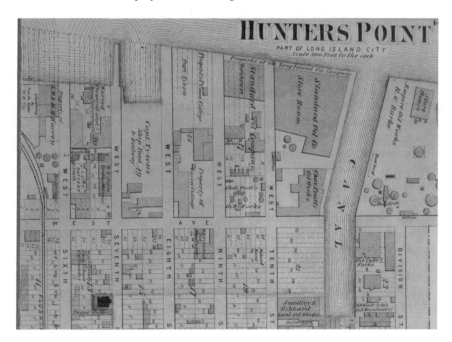

Fig. 9.1 The Standard Oil Company and the Chas Pratt Oil Works sites are indicated at Hunter's Point. A strip of land along the East River is labelled "Property of the Long Island Oil Company." The map is dated 1873 (Courtesy of The New York Public Library [21])

were already part of Standard Oil. The Long Island Oil Works had been purchased by Standard Oil in 1870. Empire Oil Works, Queens Oil Works, and Franklin Oil Works had one owner whose business address (181 Pearl Street) was that of Rockefeller and Company; the Brooklyn Oil Works, Green Point, Brooklyn E. D., owner Wm. A. Byers, also held offices at 181 Pearl. (The New York Directory of 1873 gives Standard Oil Company's address as 181 Pearl Street and includes the names of officers and the company's $1,000,000 valuation. From 1874 to 1882, the company's business address was 140 Pearl Street.) The Central Oil Works, 66th Street, N(orth) River, owner Lombard Ayres & Co, 58 Pine St, may have been independent in 1872, but

Island; owner Olophine Oil Company, 322 Broadway, N.Y.; 7. Brooklyn Oil Works, Green Point, Brooklyn E. D.; owner Wm. A. Byers, 181 Pearl St., N.Y.; 8. Central Oil Works, 66th Street, N(orth) River; owner Lombard Ayres & Co, 58 Pine St., N.Y.; 9. Hudson River Oil Works, Bulls Ferry, NJ; owner I. W. Wickes, 120 Maiden Lane, N.Y.; 10. Locust Hill Oil Works, Newtown Creek, Long Island City; owner I. Donald & Co. 124 Maiden Lane, N.Y.; 11. Union Oil Works, Brooklyn E. D. owner T. Meyers, 126 Maiden Lane, N.Y.; 12. Washington Oil Works, Newtown Creek, Brooklyn E. D. owner Thomas McGoey, 143 Maiden Lane, N.Y.; 13. Wallabout Oil Works, Brooklyn E. D. owner S. Jenney & Son, Kent Ave. foot of Rush Street, Brooklyn E. D.; 14. Vesta Oil Works, Gowanus Creek, Brooklyn; owner W. and G. F. Gregory, 125 Maiden Lane, N.Y.; Geo. Summer I, Corner Warren and 1st St, Jersey City; 15. Peerless Works, Brooklyn S. D. foot of 25th Street. Owner Denslow & Bush, 128 Maiden Lane, N.Y.; 16. Long Island Oil Works, Long Island City; offices at 140 Pearl Street, N. Y. Owner Long Island Oil Works, 140 Pearl Street, N.Y.

by 1875 Jabez Bostwick identified that Hudson River and West 66th Street location as one of his business addresses.

In 1873, the country endured a prolonged economic recession [22]. Among the players in petroleum business, the Standard Oil Company was the closest thing to a Swiss bank. A closely-held shareholder-owned company, Standard could offer both cash and shares in the company to induce owners of refineries in Ohio, Pennsylvania, and New York to sell their refineries to Standard Oil.

In this period of expansion, two New York refineries were particularly attractive to Standard Oil: the Sone and Fleming Refinery (Kings County Oil Works) held a large area of Newtown Creek waterfront on the Brooklyn side, and the Charles Pratt Refinery, which held real estate along the Queens shore of Newtown Creek and along the East River in Williamsburg, was celebrated for its premium product (Astral Oil) and recognized for its excellent refining technique. Henry Rogers, Charles Pratt's partner, had been a leader in the forces that defeated the South Improvement Company. Rockefeller recognized in Rogers a worthy adversary.

9.7   The Charles Pratt Refinery

At the Charles Pratt Refinery on Newtown Creek, Pratt was the business partner and Rogers the technology partner. Pratt had just set about constructing a refinery when William Rockefeller arrived in New York. As principals in closely-held refining companies, Rockefeller and Pratt shared a concern for properly distilled petroleum, although neither was himself in any sense a petroleum engineer. Rockefeller had Sam Andrews supervising operations in Cleveland, and Pratt would soon have Henry Rogers at Newtown Creek. As Standard's stealthy buying spree advanced along Newtown Creek, Pratt held firm. Pratt's business held two strong positions; each was back-up by one or more US patents. Pratt had facilities, including attractive waterfront refineries in New York harbor at Newtown Creek and in Williamsburg, patents for design and manufacture of high-quality cans, and proprietary methods for refining. The Pratt refinery was the industry's gold standard. Pratt's premium Astral Oil embodied the company's attention to refining. Henry Rogers's know-how stood behind Astral Oil's quality.

Roger's patent ("Improvement in Distilling Naphtha and Other Hydrocarbon Liquids," US Patent 120,539, October 31, 1871) was assigned to the Charles Pratt Refinery. The patent taught a method for separating naphtha components from kerosene by distillation. This method, which by isolating the more easily ignited naphtha components from kerosene, made the kerosene significantly safer. This method would provide basis for the reassuring words "will not explode" printed on each can of Astral Oil. While the language in Rogers's patent specification suggests his greater interest in isolating naphtha to support its markets, Rogers makes clear he understood the safety implication of that technology. In naming his original Cleveland refinery "the Standard," William Rockefeller sought to convey the message that the company's distilled product was held to high standards of preparation.

When Standard Oil first indicated its interest in the Pratt Company and Astral Oil, Rogers argued in favor of merging with Standard Oil as Pratt resisted. When the Pratt partners agreed to proceed and came to terms with Standard Oil in 1874,[13] both took shares in Standard Oil in lieu of cash, a decision that assured both Pratt's and Rogers's wealth. Although he became an officer of the company and maintained a continued correspondence with John Rockefeller, Pratt took an increasing interest in developing a technology-directed institution of higher education in Brooklyn. Rogers was an active officer at Standard Oil. Rogers knew and understood more about refining technology than did Pratt or Rockefeller, and, as he assumed an office in the company's lower Broadway headquarters, (Standard Oil moved its offices from Cleveland to New York in 1883), Rogers demonstrated previously untapped ability in management.

9.8 The Standard Oil Company of New York, An Increased Presence in New York City

Wm. Rockefeller, J.A. Bostwick, Benj. Brewster, O. B. Jennings, and Charles Pratt incorporated the Standard Oil Company of New York on August 1, 1882 at the New York Country Clerk's office.[14] The company's purpose was defined: "the refining of petroleum, the manufacturing of the various products thereof, the purchase of the crude material and the sale of the manufactured products thereof, the manufacture of barrels, boxes, cans and other packages in which the manufactured products may be kept or transported, the manufacture and restoration of acids and whatever other substances may be used in the manufacture of products of petroleum."

The document specifies that the five founders would be the company's first Trustees, its Capital Stock shall be divided into Fifty Thousand Shares of One Hundred Dollars each. The company's "principal place of business shall be New York City. The (*four or five words missing, possibly "refining and manufacturing shall be"*) carried on in Brooklyn, Long Island City and at other points in the State of New York." The founders certified that "A portion of the stock of said company shall be used for the purchase of various refineries and manufactories now existing in the State of New York belonging to the Standard Oil Company of Cleveland, Ohio, and other refining and manufacturing companies."

William Rockefeller's original New York office at 181 Pearl Street served as Standard Oil's first New York address. The company's address changed to 140 Pearl in 1874, to 44 Broadway in 1882, and finally to 26 Broadway in 1885. William Rockefeller, who had lived in and about New York City since 1866, established

---

[13] The dates of acquisition of various refineries by Standard Oil are difficult to determine with certainty. We include a map dated 1873 (Fig. 1) that indicates a Chas Pratt Refinery location adjacent to the Standard Oil facility at Hunters Point, near the southwestern corner of Queens County.

[14] The document chartering the Standard Oil Company of New York is filed at the Division of Old Records of the Office of the Clerk of New York County, at 31 Chambers Street, New York, NY.

a Manhattan address, 689 Fifth Avenue, in 1877. John Rockefeller moved from Cleveland to New York in 1883 and made a home at 4 West 54th Street in 1885.

9.9   From Whale Oil to Petroleum

The export of whale oil and spermaceti oil through New York harbor continued through the nineteenth century, but the volume of whale oil export tapered steadily once petroleum entered the market. As petroleum markets grew, the size and dollar values of whale oil exports seemed quaint, a vestige of a simpler time. Compared to the growing petroleum ventures, the wealth that whale markets had generated seemed modest. The new volumes and dollar values of petroleum exports suggested that the country had moved on from a more innocent time [23].

In 1850, the values of spermaceti oil and whale and other fish oil exported from the United States was $1,461,620; their volumes totaled about 2.1 million gallons. In 1859, the dollar amounts totaled $2,326,496, on volume totaling 2.3 million gallons.

In 1865, whaling was a centuries-old practice while the petroleum industry was barely six years old. In the fiscal year that ended in June, the Treasury Department's report on exports included data on exports of coal oil, crude petroleum, refined petroleum, spermaceti oil, and whale-and-other-fish oil. The traditional, the current, and the new oils competed in international markets. Exports of the mineral oils (coal oil and crude and refined petroleum) overwhelmed those of spermaceti and whale oil: over $16 million in value on volumes totaling about 25 million gallons. For that same year, the value of spermaceti and whale oil exports totaled $2.3 million dollars on volumes totaling 1.3 million gallons. 92% of the exported crude petroleum, 51% of the refined petroleum, and 90% of the spermaceti oil, sailed from New York harbor.

The export numbers for whale and fish oils and for coal oil began to fall as the export of petroleum oils soared. In 1870, the year the Standard Oil Company launched, the value of exported spermaceti, whale, and other fish oils totaled just over $1 million; exports of crude and refined petroleum exports totaled nearly $32 million. Nearly all the exported petroleum was refined oil; 60% of the refined petroleum left via New York, while Philadelphia handled 37%.

In 1880, the value of "illuminating oils" exported from the United States was $31,783,575. The industry had large-volume customers in Belgium, Germany, England, the British East Indies, Italy, Japan, the Netherlands, and the Dutch East Indies. About 73% of the exported illuminating oil left through New York harbor, and 21% through Philadelphia.

Both whaling and wildcatting presented rough romantic images of American fortune hunting. Both pursuits lured the young and ambitious to dangerous work in harsh places; both concerned the hard-won capture of crude oil and its refining for high-end applications. The wealth each pursuit generated relied on buyers in far-away markets, and both businesses relied on New York harbor for access to those markets.

The petroleum industry brought to New York, and through New York, to the world extraordinary low-cost-high-energy refined fuels. Petroleum launched a new age; it became irresistible. Only now, 160 years after Drake struck oil at Titusville, has the industrial world begun to reckon with the environmental costs of abundant petroleum.

References

1. Melville H (1851) Moby-Dick, or the whale. Harper and Brothers, New York
2. Yergin D (1991) The prize, the epic quest for oil, money and power. Simon & Schuster, New York, pp 20–21
3. Dartmouth College (2019) George Bissell, Class of 1845, Creates America's First Oil Company. https://250.dartmouth.edu/highlights/george-bissell-class-1845-creates-americas-first-oil-company, Accessed February 13, 2022
4. Anon (1853) New light – kerosene gas. Sci Am 9(4):29
5. Silliman B (1855) Report on the rock oil, or petroleum, from Venango Co. Pennsylvania. J H Benham's Steam Power Press, New Haven
6. Yergin D (1991) The prize, the epic quest for oil, money and power. Simon & Schuster, New York, p 22
7. Yergin D (1991) The prize, the epic quest for oil, money and power. Simon & Schuster, New York, p 25
8. Black B (2000) Petrolia, the landscape of America's first oil boom. The Johns Hopkins Press, Baltimore
9. Nevins A (1953) Study in power: John D. Rockefeller, industrialist and philanthropist. Charles Scribner's Sons, New York.
10. Chernow R (1998) Titan: the life of John D. Rockefeller. Random House, New York
11. Chernow R (1998) Titan: the life of John D. Rockefeller. Random House, New York, pp 76–77
12. Gesner's US patents on coal oil refining: patents numbered 11205, 12612, 12936, and 12987
13. Nevins A (1953) Study in power: John D. Rockefeller, industrialist and philanthropist. Charles Scribner's Sons, New York, p 34–36
14. Nevins A (1953) Study in power: John D. Rockefeller, industrialist and philanthropist. Charles Scribner's Sons, New York, p 37–41
15. Case Western Reserve, Encyclopedia of Cleveland History; https://case.edu/ech/articles/h/harkness-stephen-v
16. Nevins A (1953) Study in power: John D. Rockefeller, industrialist and philanthropist. Charles Scribner's Sons, New York, Nevins, p 59
17. Nevins A (1953) Study in power: John D. Rockefeller, industrialist and philanthropist. Charles Scribner's Sons, New York Nevins, p 82
18. Nevins A (1953) Study in power: John D. Rockefeller, industrialist and philanthropist. Charles Scribner's Sons, New York Nevins, p 83
19. The original document of sale of "the Long Island Oil Company of Hunter's Point in the State of New York" to the Standard Oil Company of Cleveland, dated March 9, 1870, is found in the archive of the Standard Oil Company at the Dolph Briscoe Center for American History on the campus of the University of Texas at Austin. https://briscoecenter.org/collections/american-energy-industry/
20. Whiteshot C (1905) The oil-well driller; a history of the world's greatest enterprise. Charles Whiteshot, Mannington West Virginia
21. Lionel Pincus and Princess Firyal Map Division, The New York Public Library. (1873). Hunters Point. Part of Long Island City. Retrieved from https://digitalcollections.nypl.org/items/510d47e2-634e-a3d9-e040-e00a18064a99

22. Chernow R (1998) Titan: the life of John D. Rockefeller. Random House, New York, p 160
23. The titles of the federal government's "Commerce and Navigation" reports, which were prepared annually for the Congress of the United States, varied over time. Data reported here is found in reports for 1850, 1859, 1865, 1870, and 1880

Chapter 10
Industry, Invention, and the Americans; Newtown Creek, Then and Now

10.1 Then

The chemicals and petroleum industries at Newtown Creek grew from neophyte to expert, from schoolboy to master, in the late decades of the nineteenth century. Several of the early industrialists had university training (Gesner, Silliman, Herreshoff, and Nichols) while others had little more than fundamental educations (Cooper, Pratt, Rogers, and Rockefeller). They were all inventive; each found ways to leverage the technologies of other trades to new advantage.

Their timing was propitious and their location ideal. When the production of chemicals and materials grew in size and sophistication at Newtown Creek, New York had already established itself as the country's leading center of trade. New York harbor was without close rival. As the fertilizer, kerosene, chemicals, and metals industries at Newtown Creek grew in volume and sophistication, their products were shipped from that same harbor to destinations along the country's Atlantic seaboard and across the Atlantic Ocean. In constructing new manufacturing plants at Newtown Creek, the industrial community placed itself between the abundant resources of the grand country to the West of Newtown Creek and the vigorous markets across the ocean to the East.

Two distinct ideas relate to the size and significance of the country's industrial growth-spurt: (1) even in its preindustrial period, the country exported refined fuels (export of whale oil predated export of coal oil and and petroleum); fuel refining seems characteristically American; (2) the American psyche and American industry celebrate the inventive abilities of individuals. The lone wolf is a national ideal.

To consider these themes (America as generator of fuels and Americans as lone wolf inventors) we will consider (1) data on the export of crude and refined oils from the United States and (2) the record of sustained innovation and production of one gifted engineer, as presented by a contemporary professional colleague.

P. Spellane, *Chemical and Petroleum Industries at Newtown Creek*,
History of Chemistry, https://doi.org/10.1007/978-3-031-09629-7_10

Export of Oils. Data on the dollar values of various industrial oils exported from the United States indicate the evolution and growth of these exports over the middle decades of the nineteenth century. A second indication of technological advancement is evident in the methods of manufacturing and refining two inorganic materials that are essential to industrial societies: sulfuric acid and high-purity copper. The two materials are indeed distinct (oils are organic compounds; sulphuric acid and copper metal are inorganic materials). Even our metrics of assessment of the two classes of materials are different: export data for the oils, and efficiencies and scales of production for sulfuric acid and copper metal. Both address the importance of Newtown Creek's large-volume industries.

We will consider "commerce and navigation" data[1] on exports of oils from the United States over several decades in the nineteenth century, coinciding roughly with the period examined in this narrative (Table 10.1). Note that the values of exported oils reported are those of the entire country, not solely product of Newtown Creek, but note also that most of the country's exported oil (85 to 90%) sailed from New York harbor. (After 1870, Philadelphia became a significant port of export for petroleum.) Also note that over these decades, the Treasury Department reset its reporting of exported oils as new types of oil were traded [2].

Inventor and Inventions. On a Friday evening in January 1908, the chemistry community in New York City met for dinner and the presentation of an award. Members of the Society of Industrial Chemistry, the American Chemical Society, the American Electrochemical Society, the Chemists' Club of New York City, and the Verein Deutscher Chemiker came to witness presentation of the Perkin Medal to Mr. John Brown Francis Herreshoff. For forty years, Herreshoff had managed operations at the Nichols Chemical Company, Nichols Copper Company, and General Chemical Company. The Society of Industrial Chemistry established the Perkin Medal a year before to recognize excellence in industrial chemistry on the part of an American,

---

[1] The United States Census website provides a history of the federal government's reporting on trade data [1]: From the earliest years of the republic, the Secretary of the Treasury has reported the volumes and values of goods imported into and exported from the United States. Revenue collected as "customs," duties collected on imports and exports, provided the major part of the country's operating funds. In 1820, the Congress charged the Secretary of the Treasury with preparing detailed reports on the kinds, quantities and dollar values of goods imported into or exported from the United States, including information on their ports of entry or export and the sources or destinations of imported and exported goods. Beginning in 1821, the reports prepared by local customs offices were consolidated and published in annual volumes entitled "Commerce and Navigation of the United States." The Commerce and Navigation reports provide exquisite detail on the values and quantities of goods imported into and exported from the United States and the sites of import and export. An Act of Congress approved in 1866 established the Treasury Department's Bureau of Statistics and specified that goods produced or manufactured in the country be distinguished from those that had been manufactured outside the country and later exported from the United States. An Act of Congress executed in 1903 moved the Bureau of Statistics from the Treasury Department to the new Department of Commerce and Labor. In 1923, Congress moved the Section of Customs Statistics of New York from the Treasury Department to the Department of Commerce, and by order of the Secretary of Commerce, it was consolidated with the Division of Foreign Trade Statistics of the Bureau of Foreign and Domestic Commerce. In 1941, the Division of Foreign Trade Statistics was transferred to the Bureau of the Census.

Table 10.1 The dollar values of various fuel oils exported from the United States in five selected years: 1840, 1850, 1860, 1870, and 1880[a]

	Coal oil	Petroleum, crude	Petroleum, refined		Spermaceti	Whale/fish
1840					$430,490	$1,404,984
1850					$788,794	$672,640
1860					$1,789,089	$537,547
1870	$177,137	$2,060,155	$29,864,193		$794,432	$228,278
	Mineral, crude	Naphthas, (benzine, gasoline, etc.)	Iluminating	Lubricating	Spermaceti	Whale/fish
1880	$1,927,207	$1,192,229	$31,783,575	$1,039,124	$487,004	$349,109

[a] Data presented here have been abstracted from tables published in each year's "Commerce and Navigation" report, prepared by the United States Department of the Treasury

but the Society elected to award the first Perkin Medal to an Englishman, to the individual for whom the award was named, Sir William Henry Perkin. In 1907, the first Perkin award was awarded to Henry Perkin. Travelling across the Atlantic to receive the medal, Perkin received a most enthusiastic reception, a hero's welcome, in New York.

William Perkin is a hero in the practice of industrial chemistry. He was a skilled and productive research chemist and a successful businessman. His expertise in the chemistry, preparation, and commercial applications of dye molecules derived from coal tar was established. Complex aromatic compounds could be had, for the first time in abundance, from the waste generated from coal used for heating. Perkin had helped establish an industry around the design and laboratory preparation of intensely colored, stable chemical compounds derived from coal.

Perkin's work on coal tar extracts took place in England just as kerosene production was starting in New York and the wide utility of Pennsylvania "rock oil" was being reported in New Haven. Fifty years after Perkin's work on coal tar extracts introduced the world to new and brilliant dyes, the value and potential of industrial chemistry in the United States was, in the minds of New York's community of chemists, established. The economic potential of chemistry and the significance of the American industry were established. The second Perkin awardee would be an American.

By 1909, John Brown Francis Herreshoff had already enjoyed a long association with William Nichols and the companies he founded. Herreshoff applied an intuitive sense of materials and engineering to improve efficiency and quality in the production of sulfuric acid. Even as he maintained a career-long pursuit of better and better sulfuric acid, he leveraged what appeared to be a fascination with metallurgy to establish methods for high volume production of high purity copper. Sulfuric acid and copper metal often share a common mineral source. Herreshoff's success in advancing the technologies for production of the two end products, highly concentrated pure sulfuric acid and high purity copper metal, and Nichols's in building profitable businesses around each, made it easy for the Society of Industrial Chemistry to select the first American to receive a Perkin Medal.

Professor C. F. Chandler of Columbia University presented the medal "to our distinguished brother chemist, J. B. Francis Herreshoff" and spoke at length of Herreshoff's first thirty years of industrial work with William Nichols and the several Nichols companies [3].

Herreshoff's principal achievements were in sulfuric acid production and electrolytic refining of copper metal. The two materials, each of extraordinary importance to technologically advanced societies, can be prepared from mineral sources. Chalcopyrite minerals comprise nearly equal masses of copper, iron, and sulfur. Sulfuric acid can be prepared in a facile manner from elemental sulfur. Sulfur, called brimstone in the early literature, roasts in air to form sulfur dioxide; sulfur dioxide reacts with additional oxygen under more forcing conditions to form sulfur trioxide, which can react with water to form sulfuric acid, H_2SO_4.

When brimstone, which can be had in high purity, becomes expensive, pyrite mineral is the next best source. Roasted in high temperature furnaces, the sulfide mineral reacts with oxygen in air to form sulfur dioxide, which, like the sulfur dioxide formed from brimstone, can be further oxidized to sulfur trioxide, "sulfuric anhydride," which can react with water to from sulfuric acid. Herreshoff designed and continued to refine an ore-roasting furnace which could successfully roast the low-cost mineral fines to produced sulfur dioxide. Other Herreshoff patents address the recovery and purity of sulfur dioxide gas and its catalytic conversion to sulfur trioxide. Still other patents describe methods for dissolving sulfur trioxide in relatively dilute sulfuric acid to form concentrated sulfuric acid. Even as Herreshoff refined methods for the final steps in preparation of highly concentrated sulfuric acid, he claimed in patents additional improvements in the design of the ore-roasting furnaces.

Professor Chandler spoke of the scale of acid production, specifying that [3]:

at one single establishment 100 tons of distilled sulphuric acid were produced daily from pyrites, that compared favorably with the concentrated acid made from brimstone, satisfying the demands at the time for purity, at a price at which brimstone could not profitably compete.

Reviewing Herreshoff's design of furnaces designed for roasting pyrite fines, Chandler points to Herreshoff's 1896 patent that describes the furnace's removable stirrer arms. Herreshoff's series of patents on ore-roasting furnaces describe improvements to the standard "Gilchrist & Johnson" type furnace. A central shaft would turn sweeper arms over crushed minerals forcing the pieces of ore over circular furnace shelves, as liberated sulfur dioxide gas rises and is collected at the top of the furnace. Herreshoff designed removable, easily replaceable sweeper arms and, in later patents (several of which appeared after Herreshoff received the Perkin Medal), Herreshoff described designs that enable air-cooling of the central shaft and sweeper arms, designs that further extend the working lifetime of the furnaces.

Chandler points out that when the Herreshoff design was described, there were [3]:

immense quantities of fines ore in various parts of the country that could not be sold on account of the inability to roast it, especially for making sulphuric acid, without great cost and trouble. This ore was used up very rapidly thereafter, rendering available say 3,600 tons a day of fines pyrites which hitherto had been a kind of drug in the market. When one estimates that this means the production of more than 1,500 tons of sulphuric acid a day, he can see the importance of this furnace to that industry. ... Up to this time, there have been sold 478 furnaces in the United States and 644 in Europe, making a total of 1122 furnaces.

Chandler calculates that, if all these furnaces were applied to the production of sulfuric acid and assuming each furnace to be capable of burning 3000 pounds of sulfur in 24 h, these 1122 furnaces would produce 2.5 million tons of sulfuric acid annually.

In electrolytic refining of copper, blister copper, the product of ore smelted in a reverberatory furnace, is cast to form thin long anodes, which, when immersed in an electrolytic cell and subject to an appropriate positive potential is oxidized to form copper ions. The copper ions generated at the anode migrate in the electrolytic cell to the counter electrode, the cathode, where, with great selectivity, copper ions are reduced to copper metal. The metal that is electrolytically deposited on the cathode is of very high purity. Before the Nichols company undertook electrolytic refining of copper, the method had been applied on limited scale at two sites, Chandler reports, but not well understood. Herreshoff addressed the work with characteristic focus and, in the first decade of the twentieth century constructed, on one part of the Laurel Hill site, what Chandler describes as the largest copper refinery in the world. The site's output was approximately one million pounds of copper a day, roughly one-fourth the entire world's output.

10.2 Now

The Newtown Creek story, as related in this book and described with enthusiasm by Professor Chandler at the Perkin Award dinner in 1908, is a story of invention and achievement. There is another Newtown Creek narrative.

For nearly as long as there has been industrial production at Newtown Creek, there has been response by the Creek's nearby community to the foul odors, acid sludge, and petroleum spilled into the Creek.

In 1877, the Board of Health of the City of Brooklyn published "The Newtown Creek Nuisances" [4].

Fires and explosions at Newtown Creek production sites gave nearly constant work to fire fighters and to journalists of the Brooklyn Daily Eagle[2] and the New York Times.[3] The reports are vivid in detail and many in number. It appears that, during Newtown Creek's period of intense industrial production, the journalism of refinery fires had become an artform.

---

[2] The Brooklyn Daily Eagle is searchable: The Brooklyn Daily Eagle on Newspapers.com.

[3] The New York Times is searchable with subscription: https://help.nytimes.com/hc/en-us/articles/115014772767-Archives.

In September 1978, a Coast Guard helicopter patrol identified an oil slick on Newtown Creek and surmised the possibility of pollution. That led to discovery of what would later be described as a 17-million-gallon spill that involved 52 acres of soil in Greenpoint, the northernmost part of Brooklyn. The US Environmental Protection Agency examined and reported on the Greenpoint Oil spill [5]. The advocacy group Riverkeeper has shepherded litigation on the part of the State of New York in New York State courts of law pursuing ExxonMobil's responsibility to remove the oil [6]. While the company argued that its clean-up efforts began immediately after the spill became known, ExxonMobil entered into an agreement with the New York State Department of Environmental Conservation to address the problem with renewed effort in 1990.

On September 29, 2010, the US Environmental Protection Agency added Newtown Creek to the National Priorities List; Newtown Creek became a Superfund site. Remediation[4] of a site under Superfund may last decades. Newtown Creek's process is at the start of its second decade. Superfund establishes a credible hope that the site may be restored, but it does not guarantee it. Like other actions brought through civilian or government agencies that seek redress from manufacturing entities or their successors, the remediation process at Newtown Creek is likely to face resistance from entities identified as potentially responsible parties (PRPs).

It is standard for successful companies, especially science and engineering companies, to support in-house research and development efforts, to optimize and improve their existing products and to explore new methods and markets for their inventions. Efforts at environmental remediation are treated differently; that citizen groups and producers assume adversarial positions in environmental recovery seems unquestioned. Like corporate research and development, remediation costs time and money, but producers tend not to celebrate their achievements in environmental recovery as proudly as they do their R and D efforts.

The genius and innovation of the chemical and petroleum businesses at Newtown Creek inspired me to explore the Newtown Creek story, the quality of the place that attracted talent and nurtured brilliant work. That their successor businesses have been less enthusiastic to contribute to the Creek's recovery, to envision and build a twenty-first century urban industrial center, is less inspiring. Several commercial entities that prospered at Newtown Creek in their early years are now PRPs at Newtown Creek. Some of the original companies at Newtown Creek have carried their practice forward into this century. As a group, the Newtown Creek PRPs have enviable knowledge and access to the most sophisticated oil-extraction and mining and engineering skills, talents that could be applied now at Newtown Creek.

If the several parties involved in the present recovery work at Newtown Creek framed the Superfund process as opportunity and as an expression of twenty-first century citizenship, if removing heavy-metal contaminants and polycyclic aromatic compounds from the creek beds were deemed as honorable and appropriate in this century as building homes for factory workers and endowing universities were in

---

[4] The EPA Environmental Protection Agency (EPA) maintains and provides access to information concerning Newtown Creek's Superfund standing [7].

previous centuries (as individuals and companies associated with Newtown Creek had done), if technology companies responded to the Superfund process with innovative engineering, Newtown Creek's legacy of productivity and progress would be carried forward and honored.

References

1. The United States Census Bureau, Guide to Foreign Trade Statistics. https://www.census.gov/foreign-trade/guide/sec1.html. Accessed February 13, 2022
2. The titles of the federal government's "Commerce and Navigation" reports, which were prepared annually for the Congress of the United States, varied over time. Data reported here are found in these five reports: Commerce and Navigation of the United States REPORT. No. 5. General Statement of Goods, Wares, and Merchandise of the Growth, Produce, and Manufacture of the United States, Exported, Commencing the 1st day of October, 1839, and ending on the 30th day of September, 1840; A Report from the Register of the Treasury of the Commerce and Navigation of the United States for the Year ending the 30th June, 1850 (January 1, 1851); Report from the Register of the Treasury of the Commerce and Navigation of the United States for the Year Ending June 30, 1860; Report from the Register of the Treasury of the Commerce and Navigation of the United States for the Year Ending June 30, 1870. No. 5. General Statement, by Countries, of Exports, the Growth, Produce, and Manufacture of the United States, to Foreign Countries, During the Fiscal Year Ending June 30, 1870; Annual Statements of the Chief of the Bureau of Statistics on the Commerce and Navigation of the United States for the Fiscal Year Ended June 30, 1880
3. (1908) Proceedings of the American Chemical Society for the Year 1908. Eschenbach Publishing Company, Easton, p 38
4. Fisk S (1877) The Newtown Creek nuisances, report on the industries upon the water-front at Greenpoint. In Report of the board of health, city of Brooklyn, 1875–1867. Union-Argus Book and Job Printing Establishment, Brooklyn
5. The Environmental Protection Agency (2007) Newtown Creek/Greenpoint oil spill a review of remedial progress (1979–2007) and recommendations Newtown Creek oil spill site, Brooklyn, New York. https://semspub.epa.gov/work/02/348795.pdf. Accessed 30 April 2021
6. Greenpoint oil spill on Newtown Creek. https://www.riverkeeper.org/campaigns/stop-polluters/newtown
7. United States Environmental Protection Agency, Superfund Site: Newtown Creek, Brooklyn Queens, NY. https://cumulis.epa.gov/supercpad/SiteProfiles/index.cfm?fuseaction=second.cleanup&id=0206282, Accessed February 13, 2022

Index

© The Author(s), under exclusive license to Springer Nature Switzerland AG 2022   107
P. Spellane, *Chemical and Petroleum Industries at Newtown Creek*,
History of Chemistry, https://doi.org/10.1007/978-3-031-09629-7